Cracking Digital VLSI Verification Interview
Interview Success
by
Ramdas Mozhikunnath & Robin Garg

First Edition: March 2016

Disclaimer

Every effort has been made to make this book as complete and as accurate as possible, but no warranty is implied. The authors shall have neither the liability nor the responsibility to any person or entity with respect to any loss or damages arising from the information contained in this book or other resources accompanying this book.
This book is an independent work of the authors and is not endorsed by their employers.

Foreword

Modern VLSI Designs find their place in every aspect of life. The complexity of the design increased exponentially and so did the need for verifying the design before it hits the market. The Verification methodologies matured from simple directed simulations to complex UVMs, emulations, formal with many more innovative solutions to emerge. The designs also moved from dedicated functionality chips to versatile SOCs, thereby increasing the complexity of design verification. The role of a validator in the design cycle has been and shall continue to be very prominent. The latest studies from the industry experts show that the average increase in the number of designers' year over year is reducing while that of the verification engineers is steadily increasing. The study further predicts verification engineer to designer ratio to be 75:25 by the end of 2025 and by 2050, the ratio to 95:5 in the digital design domain. A verification engineer needs to comprehend the architecture, micro architecture, design details and in addition needs to have a strong command over various verification methodologies, languages and technologies. The first hurdle for an aspirant to enter a new company is a successful interview which would have a wide spectrum of areas in the verification domain.

*This very book on cracking the interview is the **first of its kind and will be a great boon for all the aspirants**. Interview is an art and science and the authors took the utmost care to scientifically master the art of succeeding a verification interview. This book equips the reader with all the aspects what an interviewer would be looking for. Being an interviewer for over a decade, **I am impressed by the kind of questions** discussed in this book and I wish I had this book earlier. There are so many topics which I felt that I should have known earlier that would have made the interviews interesting. **I have no doubt in recommending this book to all the aspirants** and even the interviewers in the domain to get a new and vivid perspective. The gamut of the questions would benefit from a novice intern to the senior expert executives. This book touches every aspect of the verification engineer selection interview including technical and non-technical traits.*

The authors of this book, *Ramdas and Robin Garg have taken a great care in dealing with all the characteristics of a successful verification engineer. Ramdas has been known in the verification circles as one of the finest experts in the domain. He has versatile experience in various top class companies verifying the most complex designs and hosting the most followed verification courses online. Robin Garg, with his innovative and enthusiastic mind-set has conceived this philanthropic act of segregating the toughest of the questions and bring out a book to help validation aspirants. I have personally worked with* both the authors and I am excited to share my feelings as a foreword for this book. I wish the authors all the very best in their endeavour to bring out this book.

Achutha Kiran Kumar, Formal Verification Lead, Intel
Co-Author, Formal Verification: An Essential Toolkit for Modern VLSI Design

Verification is an art. It is one of the **most challenging** and exciting phase of Microprocessor development lifecycle. For a predictable time to Market and success of a VLSI product, Verification is the key.

If you are passionate to work in Digital VLSI Verification domain, often wonder what it takes to land a job as Verification Engineer, and want to know what skills are required to crack a VLSI Verification interview: **this book is a must-have for you!** Having a good understanding of VLSI Design/Verification concepts and insights into VLSI interview process will **reduce your anxiety and boost confidence**.

The authors: Ramdas M and Robin Garg are experts in Microprocessor Verification with combined experience of more than two decades. Through this book, they are giving **a tool that can be used by readers to expand their domain knowledge** and be prepared for successful interview. The book is well organized into chapters with detailed descriptions and explanations for various questions, providing readers with a guide to educate them. Book covers both technical and non-technical aspects of the VLSI verification interview process. The authors have **covered various concepts in great detail**: from basic design concepts to advanced verification methodologies.

Using this book as a tool rather than a guide for your preparations will assist you in your endeavors to have a successful career in the world of VLSI Verification!

Pushkin R Pari, Senior Staff Engineer, Qualcomm

Dedication

*To our Teachers, Parents, Mentors, Families, and Friends,
Thank You!*

Feedback

As a reader of this book, you are our most important source of feedback and hence we want to hear from you. You are a valuable critic and a reviewer for us. We want to know what we have done right, what we could do better, what all areas would you like to see us publish in, and any general feedback or comments are also welcome. Please leave your ratings, comments, and reviews. For any additional feedback, you can email us at "verif.excellence@gmail.com". Your feedback is highly appreciated!

About the Authors

Ramdas Mozhikunnath
Ramdas is an Expert Verification Engineer with a passion for continuous learning. He has more than 15 years of experience in pre-silicon and post-silicon verification of complex ASIC designs and Microprocessors. He is presently working as a Senior Verification Engineer and Manager at Applied Micro on verification of latest generation ARM server CPU designs. In his career, Ramdas has worked at several top companies including Intel, IBM, and many other start-ups that delivered successful verification projects. His area of expertise include: microprocessor cores, caches, coherency and memory sub-system microarchitecture, and verification. He has deep understanding of Verification methodologies and programming languages like SystemVerilog, C++, OVM, and UVM.
Ramdas believes in keeping technical skills on the cutting edge. Hence, he is passionate about sharing knowledge in the field of Functional Verification through online courses, offline sessions, and blogs on the website: www.verificationexcellence.in. During last two years, he has released three verification courses on Udemy that have registered nearly 10K users, and have received good feedback/reviews from several students and functional verification engineers.
You can follow Ramdas on LinkedIn || Twitter || Quora.

Robin Garg
Robin is an experienced semiconductor professional and a technology enthusiast. He graduated with Electrical and Electronics Engineering degree from BITS Pilani, India in the year 2011 and has been working in the field of Digital VLSI Verification since then. Over the span of last five years, he has worked for top semiconductor companies in India and UK, and has explored various facets and dimensions of Digital VLSI Verification at Pre-Si, Emulation and Post-Si levels.
On non-technical front: Robin is a Mentor, Volunteer, Runner and a wannabe Blogger. He has been mentoring Engineering/STEM students in association with various non-profit organizations, and has volunteered for numerous social causes. He is a fitness freak and is also leading various fitness initiatives for his Alma-mater: BITS Alumni Association.
You can follow Robin on LinkedIn || Twitter || Quora.

Note: This book is an independent work of the authors and is not endorsed by their employers.

Credits and Acknowledgements

Authors would like to thank following Leaders (in alphabetical order) for taking time out of their busy schedules to share their personal views on: *"What do they look for while interviewing candidates and how do they usually arrive at a decision if a candidate should be hired?"*
1) Chaitanya Adapa, Engineering Manager, Intel
2) Durairaghavan Kasturirangan, Principal Engineer/Manager, Qualcomm
3) Pradeep Salla, Technical Manager, Mentor Graphics
4) Roopesh Matayambath, Principal Engineer, Applied Micro

Authors would also like to thank following Technical Leaders (in alphabetical order) for taking time out of their hectic schedules to provide honest foreword/review on this book.
1) Achutha Kiran Kumar V, Formal Verification Lead, Intel. Co-Author, Formal Verification: An Essential Toolkit for Modern VLSI Design
2) Pushkin R Pari, Senior Staff Engineer, Qualcomm

Table of Contents

About the Authors 6
Preface 10
A Career in ASIC/SOC Design Verification 11
Introduction 13
Preparing for an Interview 14
    Interview Process and Latest Trends 14
    How should a Candidate prepare for an Interview? 14
    General Tips/Best Known Methods 16
What Leaders look for while Interviewing Candidates? 17
    Interview of First Verification Leader 17
    Interview of Second Verification Leader 20
    Interview of Third Verification Leader 21
    Interview of Fourth Verification Leader 23
Chapter 1: Digital Logic Design 25
    1.1 Number Systems, Arithmetic and Codes 25
    1.2 Basic Gates 28
    1.3 Combinational Logic Circuits 34
    1.4 Sequential Circuits and State Machines 38
    1.5 Other Miscellaneous Digital Design Questions 48
Chapter 2: Computer Architecture 51
Chapter 3: Programming Basics 68
    3.1 Basic Programming Concepts 68
    3.2 Object Oriented Programming Concepts 80
    3.3 Programming questions 85
        3.3.1 UNIX/Linux 85
        3.3.2 Programming in C/C++ 89
        3.3.3 Programming in PERL 96
Chapter 4: Hardware Description Languages 101
    4.1 Verilog 101
    4.2 SystemVerilog 109
Chapter 5: Fundamentals of Verification 139
Chapter 6: Verification Methodologies 150
    6.1 UVM (Universal Verification Methodology) 150

    6.2 Formal Verification 180

    6.3 Power and Clocking 183

    6.4 Coverage 190

    6.5 Assertions 199

Chapter 7: Version Control Systems 209

    7.1 General 209

    7.2 CVS 210

    7.3 GIT 212

    7.4 SVN 214

Chapter 8: Logical Reasoning/Puzzles 216

    8.1 Related to Digital Logic 216

    8.2 General Reasoning 219

    8.3 Lateral Thinking 223

Chapter 9: Non Technical and Behavioral Questions 225

Closing Remarks 228

Preface

Digital VLSI Design Verification practices have evolved and they continue to evolve rapidly. Historically, writing directed tests and simulating them against a design was a laborious and time consuming process. With exploding design complexity, verifying a design has become the most critical task and is usually the longest pole in a project schedule. This is driving many new innovations that can improve verification productivity. Hardware Description Languages (HDL) have evolved to support more Verification enabling constructs and new verification methodologies like: constrained random verification, Coverage and Assertion based Verification have become more popular. SystemVerilog language and Universal Verification Methodology (UVM) have gained wider adoption for dynamic simulations while Formal Verification continues to be used more and more for static simulations.

Due to these developments, Interview methods and hiring processes have also changed to look for the right candidate with relevant skills. Hence, whether you are a new starter aspiring for a career in Digital VLSI Verification, or a professional already working in this field, you need to have strong fundamentals, and be nimble enough to embrace new technologies, understand evolving methodologies, and adopt best practices.

Be it any field, working on new standards and methodologies is extremely challenging if you are not thorough with basic principles and fundamentals. And specifically in the field of Digital VLSI Verification, learning process is bit convoluted as there are hardly any resources that intend to cover combination of basic fundamentals, their application, and evolving methodologies. Hence, people looking for a job in Digital VLSI Verification field, often ask - *"What all resources do I have at my disposal for Interview Preparation? What all concepts do I need to brush up before an Interview? What all books/papers should I refer to?"*

Keeping this problem statement in our mind, we thought of writing a book that could act as a golden reference and provide one-stop-shop solution for all these questions by covering various topics that apply to Digital VLSI Verification Interviews: basic fundamentals, advanced concepts, evolving methodologies, latest trends, aptitude problems, and behavioral questions. Aim of this book is to help candidates test, brush-up, and hone basic fundamental concepts through question and answer approach. As authors of the book (with 20+ years of combined work experience in this domain), we have tried our best to cover concepts which we feel are applicable to digital VLSI verification domain through a good set of 500+ questions.

A Career in ASIC/SOC Design Verification

Quite often, I come across this question: **What is the career path for an ASIC verification engineer?** It's astounding to see the number of people having doubts and queries regarding career in the field of ASIC/SOC Design Verification. Even though this doubt is primarily surfaced by Students, Recent College Graduates and Junior Engineers, the number of Senior Engineers having this confusion is also equally significant. Hence, I thought of sharing my perspective on this with all our readers.

I am working as an ASIC verification engineer for more than 16 years now, and to be honest I had the same question in my mind in early stages of my career. I was literally confused. But I had some great Mentors who guided me and helped me with all my doubts and queries. I was lucky to have people around me who shared their experiences and stories with me. They motivated me and showed me the right path.

And as I look back, I see that I have learned a lot over the years. I have worked on Verification of several complex design projects at different companies with amazing technologies and with some of the best minds in the world. I still continue to enjoy a career in Verification and here are my inputs based on what I have experienced:

Over last several years, complexity of designs has increased and it continues to increase. Verifying a design is always crucial, as any functional defect in the manufactured chip is going to cost huge money in terms of a new tape out and would present the potential risk of losing a design-win opportunity in the market. My experience shows that project life-cycles are shrinking and there are always bugs to be found in lesser time for every subsequent project. Hence, new methodologies, processes, and innovations are critical and would continue to remain important. Many people feel that innovation in ASIC/SOC Design Verification is only limited to EDA tools. This is NOT true. EDA tools facilitate lots of usage models, but in reality these usage models are defined and applied by Verification Engineers and Architects.

Becoming an expert in Verification Domain is not an easy job. It involves much more than just running a test. A successful ASIC verification engineer should have good software programming skills, thorough understanding of various verification concepts (for modelling testbenches/stimulus), sound hardware/logic design reasoning skills (for understanding the internals of design micro-architecture), and good critical thinking skills (that facilitate understanding all aspects of a design and finding all the defects efficiently).

In today's SOC design world, the scope of an ASIC verification engineer has increased from mere functional simulations to Formal Verification, FPGA/prototype emulation, HW/SW Co-Verification, Performance Verification, and many more. There are career opportunities in each of these areas where you can deep-dive, build your expertise, and become a valuable asset for a company.

With all that said, you can build a career starting from a beginner level Verification Engineer to an Expert Verification Engineer who could be respected, could influence people, and could contribute towards goals of a company all the way from product definition, design architecture, SW development, and even customer deployment and interactions. I have personally seen some people doing this.

It's always good to have a mentor who can help you with all your queries and solve all your doubts regarding your career. Usually I try to have my role models as my mentors. If you are working in an organization look for the people who you think you would want to be like in next 5-10 years. Talk to them about their career progression, how they reached there, and discuss your career plans with them.

There can be different designations through this career path, and these designations may vary from company to company, but that should be given less importance. According to me, Learning should take preference. And like any other field, becoming an expert is not an easy affair. You are the one who need to put in all the hard-work to achieve success. There would be difficulties, there would be obstacles, but you need to be strong and motivated enough to move ahead. And, having clarity on different career options available for VLSI Verification Engineers would definitely help!

All the Best!
- Ramdas M

Introduction

How should I prepare for an interview? What all topics do I need to know before I turn up for an interview? What all concepts do I need to brush up? What all resources do I have at my disposal for preparation? What does an Interviewer expect in an Interview? These are few questions almost all individuals ponder upon before an interview. Keeping these questions in our minds, we decided to write a book that could act as a reference for candidates preparing for Digital VLSI Verification Interviews. Purpose of this book is NOT to list down all the questions that could be asked in an interview and overload the readers with thousands of questions. Aim of this book is to enable our readers practice and grasp important concepts that are applicable to Digital VLSI Verification domain through Question and Answer approach. Hence, while answering the questions in this book, we have not restricted ourselves just to the answer. Wherever possible, we have tried to explain underlying fundamentals and concepts.

This book consists of 500+ questions covering wide range of topics that test fundamental concepts through problem statements (a common interview practice which the authors have seen over last several years). These questions are spread across nine sections and each section consists of questions to help readers' brush-up, test, and hone fundamental concepts that form basis of Digital VLSI Verification. However, this book is NOT "just" about technical concepts. Scope of this book goes beyond just the technical part. It's true that strong technical skills are a must-have, but securing a job is not "just" about technical skillset. Behavioral skills also form a critical part of working culture of any company. Hence, this book consists of a section on behavioral interview questions.

In addition to technical and behavioral part, this book touches upon a typical interview process and gives a glimpse of latest interview trends. It also lists some general tips and Best-Known-Methods to enable our readers follow correct preparation approach from day-1 of their preparations. Knowing what an Interviewer looks for in an interviewee is always an icing on the cake as it helps a person prepare accordingly. Hence, we spoke to few leaders in the semiconductor industry and asked their personal views on "What do they look for while Interviewing candidates and how do they usually arrive at a decision if a candidate should be hired?". These people have been working in the industry from many-many years and they have interviewed lots of candidates over past several years. Hear directly from these leaders as to what they look for in candidates before hiring them.

In this Digital Era, where new technologies are evolving fast, a question-bank can never be 100% complete. It is impossible to cover ALL the concepts, however big the question-bank may be. Hence, to re-iterate: this book is our sincere effort to help our readers brush-up, test, and hone various fundamental concepts through questions which we feel are applicable to Digital VLSI Verification.

We hope that you enjoy reading this book. We are open to your feedback. Please let us know your valuable comments and reviews.

Preparing for an Interview

Interview Process and Latest Trends

Finding right candidate for a job opening is not an easy process. Each company has its own matrix and criteria for evaluating candidates. In general, Recent College Graduates are either hired as full time employees or as interns through On-Campus University recruitment programmes. In some countries, Graduate Fairs are also organized. Few VLSI companies conduct off-campus recruitment drives as well for hiring new graduates. Graduate hiring usually consists of a written test and/or resume screening, followed by few rounds of technical interviews, and finally a behavioral interview. Few companies conduct live online programming test as well.

On the other side, process of hiring experienced professionals is bit different. Experienced people either apply directly through a company's website, or through Employee Referral process. Some individuals and companies use services provided by hiring consultants as well. Nowadays, LinkedIn is used extensively by companies and hiring consultants to find suitable candidates for various job roles. As part of hiring process, experienced professionals usually submit their resumes and cover letters. Resume screening is followed by a telephonic interview, few rounds of Face to Face technical interviews (number of rounds may vary from company to company), and a behavioral interview.

In recent times, some companies have added one new step where they send a questionnaire to the Job Applicants, ask them to solve the questions, and send back the questionnaire in 1-2 hours. This step is increasingly being used by some companies as an additional step before a telephonic interview, or as a replacement for telephonic interview. Questions in these questionnaires typically checks for a candidate's ability to think through a real-life verification problem, define a verification strategy, and suggest a best possible method to solve the problem. Applicants are sometimes also asked to comment on pros and cons of using different verification tools/methodologies.

Hence, looking at interview processes and latest trends, it is recommended to have a LinkedIn profile. Active participation in various forums (*where real-life verification problems, latest verification trends, and verification advancements are discussed*) is also encouraged.

How should a Candidate prepare for an Interview?

Nature and difficulty level of questions for Digital VLSI Verification vacancies vary depending upon the experience of the candidate (interviewee) and the Job requirement. In order to make things easier and less complicated for our readers, we have followed a "Divide and Conquer" approach, where we have divided interview questions into "Nine" categories.

These nine categories (described below) cover almost all the main topics which are usually asked in a Digital VLSI verification job interview.

As we mentioned in the introduction section, we have tried our best to answer questions in a way that would help readers understand basic fundamentals, rather than just overloading them with thousands of questions.

Digital Electronics/Digital Logic: This forms very important part of any Digital VLSI Verification Job Interview as this is the topic where most of the engineering knowledge applies. Wide range of questions related to digital logic may be asked. In this book, we have subdivided this section into Number Systems, Arithmetic and Codes, Basic Circuits, Combinational Logic Circuits, Sequential Circuits and State Machines, and have also listed some miscellaneous Digital Design questions.

Computer Architecture: With VLSI designs trending towards System on Chip (SOC) designs, understanding fundamentals of computer architecture have gained a lot of importance for VLSI Design Verification. In this section, we cover questions on processor architecture, memories, caches, and instruction sets.

Programming Basics: Present day Design Verification job is visibly becoming more and more software oriented and good programming skills are necessary for any testbench development and/or Verification Intellectual Property (VIP) development. This section test candidate's programming fundamentals through questions related to basic programming languages like C/C++, Perl, Shell, OOP concepts (Object Oriented Programming), and UNIX/Linux fundamentals.

Hardware Description Languages: Hardware Description Languages (HDL) are special type of programming languages that are used to model behavior of digital logic circuits independent of any underlying implementation technology. In this section, we extensively discuss various fundamentals and questions relating to Verilog and SystemVerilog which forms a significant part of interview process.

Key Verification Concepts: This section consists of questions related to fundamentals of Functional Verification and how candidate thinks through a problem with a Verification mind-set. We list simple digital designs with corresponding design specifications, and ask the reader to: define a verification strategy, explain the steps to verify the design and how to identify and ensure all scenarios are verified. Additionally, this section also contains some commonly asked questions related to fundamentals of Verification. A Recent College Graduate may not be asked a lot of questions from this section, whereas this section may constitute a significant portion of interview for a senior candidate. In an Interview, difficulty of this section usually varies with the experience of the candidate.

Verification Methodologies: With plenty of tools and techniques available for the Functional Verification process, it's quite possible to deviate from the focus-area and lose track. Also, without defining a correct verification methodology, user might end up adding more complexity to the verification process, rather than finding the efficient methodology. Hence, defining an efficient methodology is part of verification planning phase. The Verification methodology can include: Dynamic Simulation vs Formal Verification, Assertion based Verification, Coverage methodology, Power Aware Simulations, Performance

Verification, and also UVM (Universal Verification Methodology) for constrained random testbenches. This section is organized into subsections that will help you understand each of these methodologies separately.

Version Control Systems: Version Control Systems have been an integral part of Software Engineering domain for a long-long time. But now they are popular in Hardware Engineering domain as well. With Hardware Designs becoming complex, various new features getting integrated every quarter, multiple folks working on the same database across different sites and locations, version control systems have become indispensable. Hence, this section touches upon basics of various Version Control Systems: CVS, GIT and SVN.

Logical Reasoning/Aptitude: This section aims at testing aptitude and logical problem solving skills. Sometimes, aptitude and programming questions are combined and candidates are asked to solve a aptitude question using a program. We have further divided questions in this section into three categories: Related to Digital Logic, General Reasoning, and Lateral Thinking.

Behavioral: This is an important part of any interview. Having strong technical skills alone won't suffice as you are usually expected to work as part of a team. Behavioral skills form an important part of work culture of any company and hence this section is usually taken very seriously by recruiters. Performing great in this section with poor technical skills may not fetch candidate a job but performing bad in this section can cost candidate a job.

General Tips/Best Known Methods

Before we move onto our next section and start with technical interview questions, it is very important to follow a correct approach from the time (Day-1) you start preparing for an interview. We have grouped together few tips and Best-Known-Methods that should be followed once you start preparing for an interview. Here are some of them:

1) Be thorough on all the skills and projects you mention in your resume. Interviewers do spend time figuring out if you actually know what you claim to know.
2) Never add fake or frivolous data in your resume.
3) It is recommended that Recent and New College Graduates brush-up on their Engineering basics, especially all the projects and the thesis (if applicable) they completed.
4) Avoid saying something like "I read this N number of years ago, hence I don't remember this". If it's there on your resume, you are accountable for it. If you don't remember it, remove it from your resume.
5) Never try to trick the interviewer. They are smart.
6) Be Honest and forthright. If you don't know answer to a question, it's better to say "Sorry, I don't know this", rather than trying to build a wrong story.
7) Don't be Nervous. Be Calm.

What Leaders look for while Interviewing Candidates?

One of the most important part for acing and cracking any interview is to know what interviewers look for while hiring a candidate. Hence, for the benefit of our readers, we thought of adding this unique and amazing section. We spoke to few industry leaders to understand their thought process and decode an interview from a recruiter's perspective. So brace yourselves, and enjoy reading candid views of few industry leaders on "**What do they look for while Interviewing candidates and how do they usually arrive at a decision if a candidate should be hired?**". Do spend some time going through this section as it would give you a very good overview on what skill sets and behaviors are quintessential.

Interview of First Verification Leader
Chaitanya Adapa, Engineering Manager, Intel

Q1. What do you look for in candidates while interviewing for Digital VLSI verification roles at different levels (say beginner, junior, mid-senior and senior levels)?

    Beginner: Basic logic design, Fundamentals of Computer Architecture, Basics of HDL (Verilog/SystemVerilog etc.), Problem Solving, Ability to learn/apply and soft skills like team orientation, communication, etc. Scripting, Software experience, validation knowledge and domain knowledge (GPU, CPU etc.) are a plus. Since the candidate at this level may not have any major experience to show the focus is on the aptitude and attitude. The person should gel with the team he is going to join and also be able to learn and apply his knowledge independently. We look for a growth mind-set where the candidate is ready to learn.

    Junior (2-3 years): The above for beginner is the basic requirement for Junior. On top of it we focus mainly on how well they understood the projects they have worked on in the last year or so. They need to be clear on what the project was, what their contributions were, how they see themselves contributing to the previous Organization and also how inquisitive are they to learn adjacent areas. The candidate should be able to convince the interviewer on all the achievements listed in the resume as the selection is based on resume. We require Validation experience for a junior. In the interview we also assess the debug capability and scripting ability. Software experience is a plus. The candidate can narrate situations where he/she has demonstrated sound technical problem solving.

    Mid-Senior (5-8 years): The above for Junior is a basic requirement for Mid-senior. At this stage in the career we will consider the candidate for a Lead position. So the candidate should demonstrate Domain expertise and concrete deliverables in Validation domain, which can include meeting or beating timelines on critical projects, presenting in conferences, patents, tools/flows/methodology etc. Should have planning and execution as a strength. It is good for a candidate to be aware of Project Management at this stage. Stakeholder Management also becomes important at this stage. Can narrate examples of how well he/she was able to plan, execute, and manage stakeholders. This involves identifying risks and finding ways to mitigate them. The candidate should be able to deliver cleanly on any challenge/problem in validation.

    Senior (9-15 years): The above for mid-senior is a basic require for a Senior Validation Engineer. At this stage of the career the candidate is expected to have significant impact on projects, timelines, design, validation or even architecture. The candidate should start identifying the problems in validation domain and propose ways to address this via design, validation collateral, process, tool, methodology or resources. Innovation,

Stakeholder Management, Influencing and Negotiation become very important. The candidate may be expected to manage or lead a team so Team building is also critical.

Q2. How do you evaluate technical skills?

The interviews are usually 45 mins to 1 hr each with 4 or 5 interviews for mid-seniors. For seniors there may be multiple interviews with multiple levels of Management. Up-to mid-seniors the interviews are more on technical. The focus on Soft Skills and leadership questions increase with the seniority.

There are questions on language syntax (System Verilog, C, C++, Perl, and Python etc.), problem solving using validation techniques (BFMs, checkers, coverage) and fundamental concepts (Boolean logic, OOP, digital circuits, caches/memory, pipelining, etc.)

For coding the candidate could be asked to write code for a specific sorting algorithm, demonstrate inheritance.

For Logic design solving Boolean logic expression, digital circuit for solving a logic problem, Multi-Cycle-Path, latency reduction techniques etc.

Q3. How do you evaluate behavioral skills?

There are multiple aspects of behavioral skills which are evaluated. Some of them are Integrity, Honesty, Team orientation, Conflict resolution, Negotiation, Stakeholder Management and others.

It is important for the candidate to be honest and sincere about the responses. The interviewer for behavioral is experienced enough to figure out any insincere comments.

Integrity is about being disciplined and professional when dealing with deliverables, customers or tough situations. If the candidate is found to make compromises or shortcuts at the cost of project, this will work against the candidate. It is important to understand the importance of delivering to what is assigned/committed.

If there is lack of experience in certain areas it is better to make it clear to the interviewer. The candidate should not be found misrepresenting experience or data during the interview.

Other aspect of behavioral is about how committed the candidate is about the interview and if he/she will eventually join if the offer is extended.

The evaluation is done using situation based questions. The candidate is requested to put him/her in a situation to determine the actions that will be taken or opinion on the outcome.

Passion for the company is also a factor. A strong passion for the company may show motivation for the candidate to put in extra effort at work.

The reason to leave the previous company is also a key factor. The reason to leave should be very convincing. If the candidate is not sure it will show up in the interview.

Q4. What's the decision process (behind the scenes) while deciding to make an offer?

Technical feedback, relevant experience for the position and behavioral feedback are considered.

Usually multiple candidates are evaluated so there is a comparison done on the interview performance of candidates.

The one closest to the requirements will be selected.

Each candidate has a set of skills they possess for the role and a few that they may lack. So the gap in skills is also a factor. Some skills are easy to ramp-up during the job. Other skills are a must and carry more weightage when deciding about the candidate.

Since there is a short time for the interviewer to evaluate the candidate, it is important the candidate create an impression during the first few minutes of the interview and carry forward the impression while answering questions. This is found to influence the interviewer in most cases.

In order to have a positive influence on interviewer it is important the candidate prepares well on his resume, has a clear objective of what he wants from a role, why he/she wants to join this company, why he is leaving the previous company.

Ideally a candidate should not be a disgruntled employee as this may indicate the same may repeat with the new company too. So the reason for leaving the previous company is very important in establishing the positive attitude of the candidate.

Q5. According to you, what are the must have skills (both technical and behavioral) for a Digital VLSI verification role?

Technical :: Logic Design, Validation, Computer Architecture, Software, Domain knowledge (GPU/CPU/Server/IOT etc.), Debug capability, Scripting, Methodology/tools/flows (SystemVerilog, Verilog etc.), Participation in Conferences, etc.
Soft-skills :: Stakeholder Management, Team Orientation, Leadership, Project Management, Negotiation, Team building, Innovation, Collaboration, etc.

Q6. If I have to ask you your favorite interview question, what would it be?

1. Why do you prefer the company you are interviewing at?
2. This will open up a discussion that can help evaluate the candidate on the seriousness for the role, commitment to the job and any insecurity on his current job.

Q7. Intern conversion to full time employee. Parameters you consider before conversion?

The parameters are similar - Technical, Experience in relevant domain and soft skills.
However since there is opportunity to observe and train the intern for much longer time the evaluation is a lot thorough.
The intern should have picked up key skills for the group he/she is interning with, delivered to the tasks assigned to him/her, showed ability to learn about the bigger picture for the project; able to work in a team, developed a network within the group.
It is important the intern learn about opportunities from time to time by meeting with the supervisor/Lead/Manager.
When it comes to deciding between interns usually delivery to tasks and maturity in domain plays a key role.

Q8. If you have an option to hire between: "Candidate with High Aptitude" and "Candidate with Future Promise" - Which one would you choose and why?

Ideally I would like both. But more often than not most of them have either one or the other and not both.
In such cases I would choose a candidate with Future Promise. One reason for this is Aptitude can be learnt but Attitude is difficult to learn or develop. So I have a better chance of success with a candidate who has good attitude, able to pick new areas and is willing to work hard towards his/her growth.

I also clearly state to the candidate on the gaps in aptitude and the reason for selection. This instils a lot of confidence in the candidate and also states the gaps the candidate needs to fill on the job.

Interview of Second Verification Leader
Durairaghavan Kasturirangan, Principal Engineer/Manager, Qualcomm

Q1. What do you look for in candidates while interviewing for Digital VLSI verification roles at different levels (say beginner, junior, mid-senior and senior levels)?

Beginner/Junior
1. Digital design knowledge – focus more on problem solving than programming skills.
2. Problem solving – Give a problem/puzzle to solve in stipulated time. Ask the person to explain the approach. Weightage is given to the approach as much as to the solution.
3. Communication – Clarity in communicating than testing the language skills. This is absolutely important because as verification engineers we may have to communicate an issue/bug over an e-mail/bug report or through a meeting across many geos and it's imperative the person can say what he thinks, not minding the language semantics.
4. Sometimes I draw a simple block diagram with few modules like CPU, memory, or any known peripheral and write some specs on how all this works. Will ask the person to come up with:
    a. 5 simple test plan items covering the given spec.
    b. 5 medium complexity Test Plan items.
    c. 2 corner case line items (typically timing sensitive paths)

Mid senior/Senior
1. All of the above plus go in depth on resume for the projects the person has owned. This is to ensure how much in-depth the person knows on what he/she did.
2. Pick 1 or 2 of the projects and ask some abstract questions to understand how much of big picture a person understands or has learnt. For Ex: understand what his product does, what's the market, who are the market leaders.
3. Give a problem statement and ask the person to come up with a complete verification TB with HLTP (High Level TP). In this process, candidates' breadth knowledge on verification can be tested.

Q2. How do you evaluate technical skills?

Approach to the question asked will be given high weightage. From the clarifying questions he/she asks, interviewer can access the person's understanding ability. This is the most important facet since verification is all about not believing, rather not assuming anything.

Q3. How do you evaluate behavioral skills?

Two important behavioral traits absolutely important for Verification engineers:

1. Communication (written/verbal) and listening skills – These can be evaluated during the interview process. Making a person go to whiteboard will gauge the person's

confidence in his subject and also we can measure how well he communicates what he thinks.
2. If the person jumps to conclusion fast for a question asked by you, he is not going to be a good verification engineer according to me. Instead he has to ask lot of clarifying questions, not assume anything and then go ahead with the problem solving.

Q4. According to you, what are the must have skills (both technical and behavioral) for a Digital VLSI verification role?

Same as the ones mentioned in Questions 2 and 3.

Q5. Intern conversion to full time employee. Parameters you consider before conversion?

1. The obvious one, How well the intern does the task assigned?
2. How well he/she works with the team?
3. Is the intern showing eagerness to learn new things?
4. How the intern adheres to the timeline?
5. And last but not the least, integrity of intern.

Q6. If you have an option to hire between: "Candidate with High Aptitude" and "Candidate with Future Promise" - Which one would you choose and why?

Depending on the need, I would need a mix of both.

Interview of Third Verification Leader
Pradeep Salla, Technical Manager, Mentor Graphics

Q1. What do you look for in candidates while interviewing for Digital VLSI verification roles at different levels (say beginner, junior, mid-senior and senior levels)?

The primary criterion is to see if the candidate fits the current job requirement. The common denominators across all levels are learning ability, problem solving skills, soft skills and technical competence.

Q2. How do you evaluate technical skills?

The technical skills are evaluated using a set of 3 – 4 technical interviews where the interviewers test the following:
1) Generic VLSI and ASIC Verification skills
2) Language skills such as Verilog, VHDL, SystemVerilog
3) Methodologies like UVM.
4) Contribution to previous projects

Q3. How do you evaluate behavioral skills?

In addition to the technical interviews, we seek help from HR for behavioural skills. Also, while we interview for Application Engineering role, behavioural skills are very important and we consider the following during the interview:
1) Body language
2) Communication skills
3) Clarity in communication

Q4. What's the decision process (behind the scenes) while deciding to make an offer?

Get feedback from all the technical interviewers and HR and consolidate their individual ratings and we consider the potential fit of the candidate for the role.

Q5. According to you, what are the must have skills (both technical and behavioral) for a Digital VLSI verification role?

1) Basics of VLSI
2) Digital Logic
3) Verilog/VHDL
4) SystemVerilog
5) C
6) Problem Solving/Debug skills
7) Good Communication skills
8) Team Player

Q6. If I have to ask you your favorite interview question, what would it be?

Convert a 2:1 Mux into NAND/NOR gate. This is good and a basic question to get started on the technical questions.

Q7. Intern conversion to full time employee. Parameters you consider before conversion?

We don't hire interns at Mentor Graphics Sales team in India.

Q8. If you have an option to hire between: "Candidate with High Aptitude" and "Candidate with Future Promise" - Which one would you choose and why?

It would depend on the need of the hour before we decide. We always look for future stars.

Q9. What will be one key message to people aspiring for great career in verification industry?

Methodology is just not sufficient, what you need is domain expertise to be successful in the verification industry.
Technical Skills + Domain Expertise = Success in Verification industry

Interview of Fourth Verification Leader
Roopesh Matayambath, Principal Engineer, Applied Micro

Q1. What do you look for in candidates while interviewing for Digital VLSI verification roles at different levels (say beginner, junior, mid-senior and senior levels)?

    Beginner: Digital Electronics Knowledge, Problem solving skills, and coding skills.
    Junior/Mid-Senior: All the points mentioned above for Beginner. On top of that, challenges they came across in their previous projects and the solutions they applied. Lots of questions on verification methodologies and verification languages they have used (For Example: OVM, UVM and SV etc.)
    Senior: All the things mentioned above for Junior/Mid-Senior. Additionally, probe on the ability to mentor juniors and skills on leading a team.

Q2. How do you evaluate technical skills?

I provide a design unit and ask them to come up with various test scenarios to verify that unit. Also, I ask general verification questions (For Example: verification of an async FIFO)

Q3. How do you evaluate behavioral skills?

I observe the way a candidate approaches conflicting questions, candidate's body language and general attitude. Also, I ask general behavioral interview questions like: "What steps would you take if there is a conflict in the team?"

Q4. What's the decision process (behind the scenes) while deciding to make an offer?

Key thing should be to evaluate what exactly the new candidate brings on to the table and how much would it be useful for the organisation. Cost to the company for a candidate can also be taken into consideration.

Q5. According to you, what are the must have skills (both technical and behavioral) for a Digital VLSI verification role?

Good Analytical skills, Debugging and Problem solving skills, Patience and Learning skills, and skills to Adapt to different requirements.

Q6. If I have to ask you your favorite interview question, what would it be?

What are the different challenges in functional verification?

Q7. Intern conversion to full time employee. Parameters you consider before conversion?

How fast a candidate can learn new things? What are the candidate's immediate goals and would the candidate be going to stay with the organisation for a longer duration.

Q8. If you have an option to hire between: "Candidate with High Aptitude" and "Candidate with Future Promise" - Which one would you choose and why?

Candidate with future promise, because it is always important to look for the future needs of the person and also what is the expectation from this person of the organisation in the future.

Q9. What will be one key message to people aspiring for great career in verification industry?

Keep on developing and improving debugging skills because this is the area where most of the time is being spent during the process of verification.

Chapter 1: Digital Logic Design

Understanding the fundamentals of Digital logic design is an essential skill for performing any job in the VLSI industry. Hence, irrespective of whether the interview is for an ASIC design job, or verification job, or any backend design or layout job, questions that test logic design skills are an important part of the interview. Hence, this is the most fundamental and most important topic for securing a job in VLSI industry. This section lists down some of the most commonly asked questions in the interviews with answers and detailed explanation of the concepts. Once you master these concepts through these questions, same concepts and logical approach could be used for related questions.

1.1 Number Systems, Arithmetic and Codes

Number systems form the basis for conveying and quantifying information in a digital system. This section consists of questions related to common number systems like decimal, binary, octal and hexadecimal (hex), arithmetic operations in different number systems, conversion between different representations etc.

> **1. Convert following decimal numbers in signed binary, octal and hexadecimal numbers using minimum possible number of bits.**
> **a) 17 b) -17**

A decimal number consists of decimal digits (0 to 9), a binary number consists of binary digits (0, 1), octal number consists of octal digits (0 to 7), and a hexadecimal number consists of 16 digits (0 to 9, A, B, C, D, E and F).
A decimal digit can be converted to any other base by following three simple steps mentioned below:
- Divide the decimal number by the base (i.e. 2 for binary, 8 for octal, and 16 for Hex)
- The remainder will form the lowest order digit in the converted number system.
- Repeat steps 1 and 2 mentioned above until no divisor remains

The MSB represents "sign" information. For negative numbers, MSB is one and for positive numbers, MSB is zero.

Using the steps mentioned above, we will get a binary value of 10001, octal value of 21 and hexadecimal value of 0x11 for decimal number 17. Since this question specifically asks for signed representation and as we have seen that positive numbers have MSB as 0, binary number "10001" represents a negative number in signed representation. Hence, we can say that 5 bits are insufficient to represent +17 in signed binary number system (as a matter of fact, 5 bits can represent decimal numbers from -16 to 15 only). Therefore, we need 6 bits to represent +17. This means that we need to append one more "0 " at the beginning of 10001. This would give us 010001 (a 6 bit number), as binary representation of +17.

One easy way to convert a positive number to a negative number in binary representation is to take 1's compliment and add 1 (which is same as 2's compliment).
Also, once we have a binary number, we can easily convert it into octal (grouping three binary bits starting from LSB) or hexadecimal (grouping four binary bits starting from LSB)

Therefore for -17,
- 1's compliment of +17 (010001) would be "101110",
- And adding 1 to above number would give us "101111"

a) Based on this, the decimal number 17 would be represented as:
   Binary = **010001**, Octal = **21**, Hexadecimal = **0x11**
b) And decimal number -17 would be represented as:
   Binary = **101111**, Octal = **57**, Hexadecimal = **0x2f**

2. What is the decimal equivalent of hex number 0x3A?

To convert a number from a non-decimal base to a decimal base, following steps are required:
- Start from the least significant digit, and move towards most significant digit.
- Multiply each digit with "<base> to the power of that <bit position>", i.e. $<base>^{<bit\ position>}$
- Sum the results over each digit.

Therefore: $0x3A = [0xA * 16^0] + [0x3 * 16^1] = [10*1 + 3*16] =$ **58**

3. What is Gray code and what are some of the benefits of using Gray code over Binary code?

A Gray code is a binary number system in which two successive values differ only in one bit. It is also known as reflected binary code

Following table shows the Gray and Binary code for values from 0 to 7

Decimal	Binary	Gray
0	000	000
1	001	001
2	010	011
3	011	010
4	100	110
5	101	111
6	110	101
7	111	100

In Binary code, a transition between two values could have transition on more than two bits and this could sometimes lead to ambiguity if different bits take different time to transition. For example: transitioning from 3 to 4 in binary (011 to 100) requires all the bits to toggle. This can lead to some intermediate values if say the three bits have different switching time.

Whereas in Gray code, since only one bit changing any time, there is no possibility of any such ambiguity.
One more advantage of using Gray code is: since fewer bits are toggling in Gray code, a design using Gray code would consume less power compared to one using a binary code.

4. What is a parity bit and how is it computed?

A parity bit is a bit which is added at the end of the string of a binary code, and it indicates whether the number of bits having a value "one" in the string is even or odd. Accordingly, there are two variants of parity code - even parity and odd parity.

To compute parity bit, the total number of bits having a value "one" in a binary code is counted. If number of "ones" are odd and if we use an even parity, then the parity bit is set to 1 so that the total number of ones including parity bit counts to an even number. If number of "ones" are odd and if we use an odd parity, then the parity bit is set to 0 so that total number of ones including parity bit counts to an odd number.
Parity bit is computed by taking XOR of all the bits in the binary string. Parity bit is used as the simplest form of error detecting code.

5. For a given binary string: 111001, compute the proper odd parity bit.

The given binary string: 111001, has four "1's". Using an odd parity, the total number of 1's in the binary string including the parity bit needs to be odd. Hence, the odd parity bit for this string will be **1**.

6. What are 1's complement and 2's complement? Where are they used?

If all bits in a binary number are inverted by changing each 1 to 0 and each 0 to 1, the resulting binary number is called the 1's complement.
For example: The one's complement of a binary number 110010 is 001101

The 2's complement of a binary number is obtained by adding a 1 to the one's complement of the number.
For example: The two's complement of a binary number 110010 is 001101+1 = 001110

The two's complement of a number is used to represent signed binary numbers. It is also used for subtraction of binary numbers. The one's complement is an intermediate step to get to two's complement.

7. What is a BCD code and how is it different from binary code? What will be the BCD and binary code for decimal number 27?

BCD is Binary coded decimal and is a four bit binary code that can be used to represent any decimal digit (from 0 to 9). A binary code is a binary representation of the decimal number,

and the number of bits needed for a binary code would depend on the decimal number. For decimal numbers 0 to 9, both BCD and binary code would be same.
A number 27 can be represented in BCD by using four bit code for both 2 and 7.
Hence, **BCD for 27 will be 0010 0111**, while the **binary code for 27 will be 11011**

8. **Which of the following code can represent numbers, characters, and special characters?**
   a. BCD
   b. Gray
   c. EBCDIC code
   d. ASCII code
   e. Alphanumeric code

e) Alphanumeric code.
Alphanumeric code is a combination of alphabetic and numeric characters and can be used for representing numbers as well as character(s)/special character(s).

1.2 Basic Gates

9. **Which of the following gates is called a universal gate and why?**
   a. AND
   b. NAND
   c. OR
   d. NOR
   e. XOR

A universal gate is a gate which can implement any Boolean function without need to use any other gate type. The NAND and NOR gates are universal gates.

10. **How can you implement a two input AND, two input OR and a single input NOT gate using two input NAND gates(s)?**

If A and B are the two inputs of a NAND gate, then the output equation for a NAND gate will be, Y = (A.B)'

1) **NOT Gate**: The NOT gate can be described with the output equation as Y= (A)'. So, if inputs A and B of a NAND gate are connected together to same input, then we get a NOT gate.

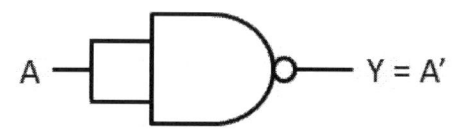

$$A \longrightarrow \boxed{} \!\!\!\!\!\!\!\!\!\!\!\!\!\!\!\! \longrightarrow Y = A'$$

2) **AND Gate**: An AND gate is described by the output equation Y= A.B and hence connecting a NOT gate to the output of NAND gate will implement an AND gate.

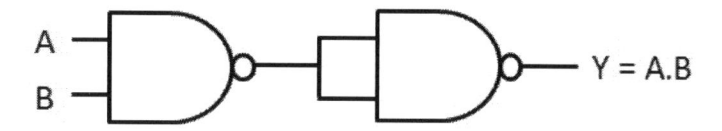

$$\begin{array}{c} A \\ B \end{array} \longrightarrow \text{NAND} \longrightarrow \text{NOT} \longrightarrow Y = A.B$$

3) **OR Gate**: OR gate can be described as Y= A+B = (A')' + (B')' = (A' B')' using De-Morgan's law. Since A' and B' can be represented as NOT gates, above equation can be represented as below which gives an OR gate.

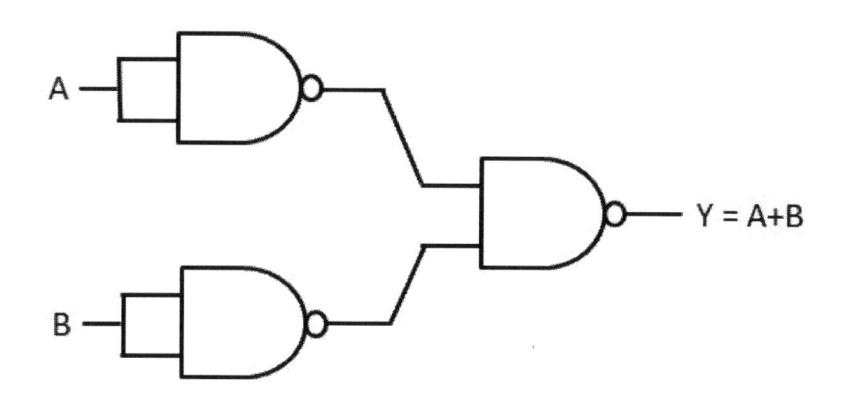

$$\begin{array}{c} A \longrightarrow \text{NOT} \\ \qquad\qquad\qquad \longrightarrow \text{NAND} \longrightarrow Y = A+B \\ B \longrightarrow \text{NOT} \end{array}$$

11. How can you implement a two input AND, two input OR and a single input NOT gate using two input NOR gates(s)?

A NOR gate is described using the equation: Y = (A+B)'.

1) **NOT Gate**: A NOT gate is described using Y = A'. If both inputs of a NOR gate are tied together Y = (A+A)' = A'

29

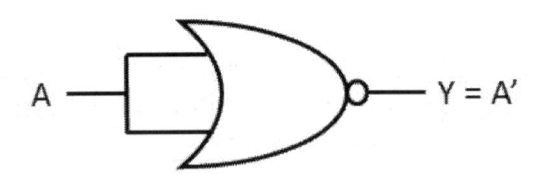

$$A \longrightarrow \boxed{}\!\!\!\supset\!\!\circ \longrightarrow Y = A'$$

2) **AND Gate**: An AND gate is described by equation: Y = A.B = (A')'.(B')' = (A' + B')'. Hence, we can use a NOR gate with inputs as A' and B'. A' and B' can be implemented using a NOT gate for A and B as shown below.

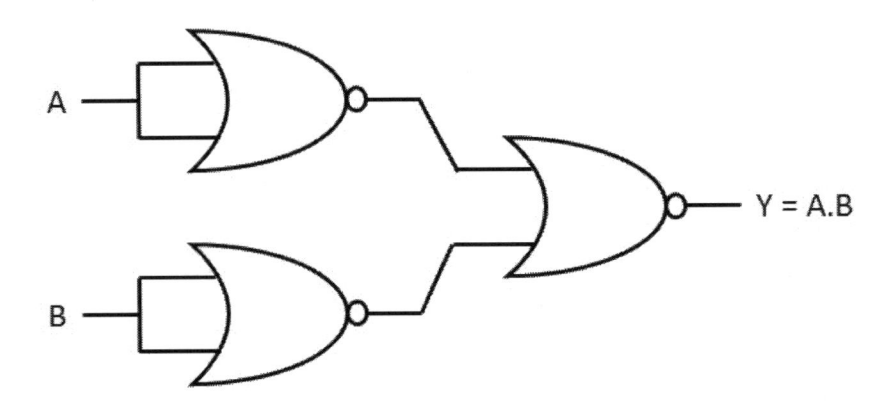

$$Y = A.B$$

3) **OR Gate**: OR gate is described by Y = A+B = ((A+B)')'. Hence, (A+B)' can be implemented using a NOR gate and the inversion of same can be done using a follow on NOT gate as shown below

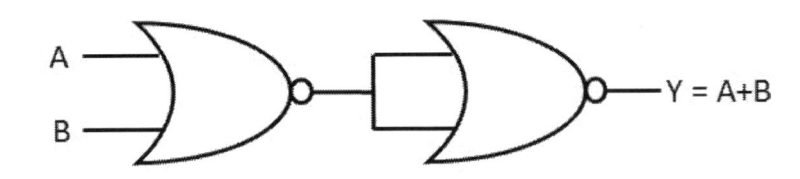

$$Y = A+B$$

12. How can you implement following gates using a 2:1 MUX?
   a) Single Input NOT
   b) Two Input AND
   c) Two Input OR
   d) Two Input NOR
   e) Two Input NAND

f) Two Input XOR

A multiplexer is a combinational logic that can multiplex two inputs onto a single output lane using select lines. A 2:1 multiplexer will have a single select input and based on the value (0 or 1), it would drive the output with either of the inputs.

a) NOT Gate: A NOT can be implemented as shown below by connecting the input of NOT gate to select line and the inputs tied to 1 and 0 as shown below.

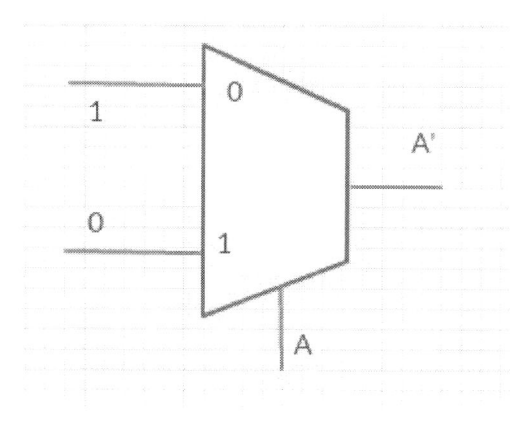

b) AND Gate: An AND gate can be implemented using MUX as shown below

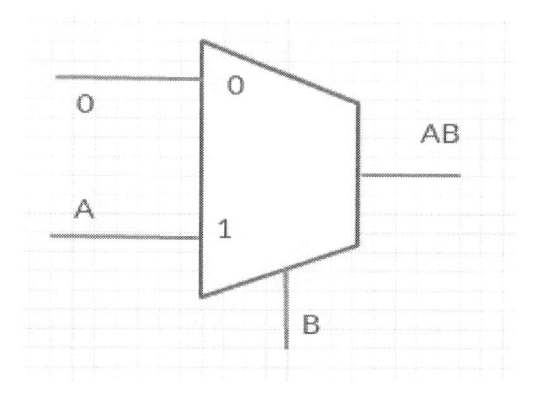

c) OR Gate: OR gate can be implemented using a 2:1 MUX as shown below.

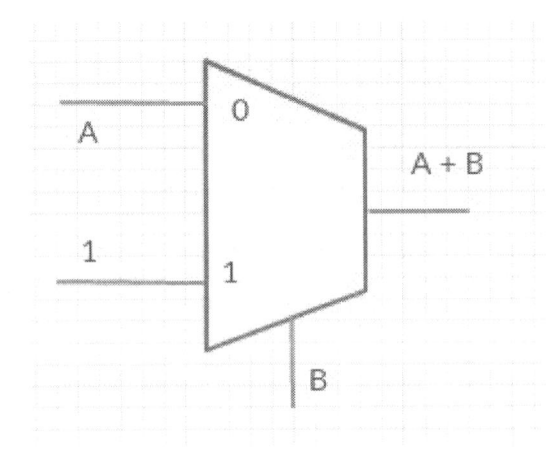

[OR gate MUX diagram: input 0 = A, input 1 = 1, select = B, output = A + B]

d) **NOR Gate:** A NOR gate can be implemented using a combination of OR gate and NOT gate from above.

e) **NAND Gate:** A NAND Gate can be implemented using a combination of the AND gate and NOT gate from above.

f) **XOR Gate:** A XOR Gate can be implemented using a 2:1 MUX as shown below. The zeroth input is connected to A and the 1 input is connected to A' (Use another MUX to implement NOT of A). The MUX output will now be AB' + A'B which is same as XOR gate.

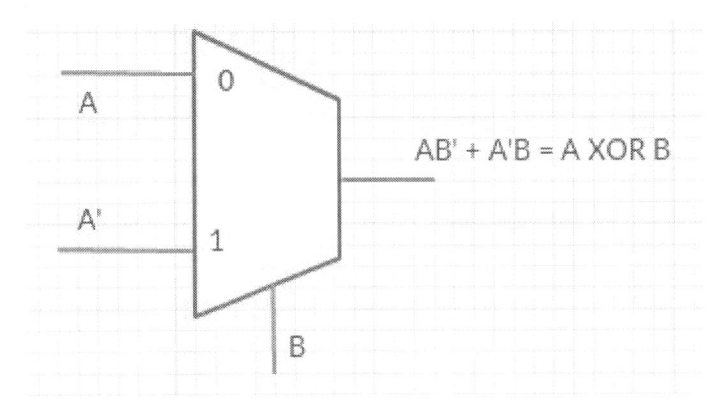

[XOR gate MUX diagram: input 0 = A, input 1 = A', select = B, output = AB' + A'B = A XOR B]

13.   What are typical uses of "XOR" gates in data communication?

An XOR gate is used in computation of error detection codes like parity, CRC and ECC. It is also used in pseudo random number generators.

14.   When can the output of a 3-input NAND gate be zero?
   a) when at least one input is zero
   b) when all inputs are zero
   c) when at least one input is one
   d) when all inputs are one

A NAND gate can get an output zero only when all inputs are 1. So, **d) is the right answer.**

15. How can you design an inverter using a XOR gate?

An XOR gate is described using the equation Y = AB' + A'B. If one of the inputs is tied to 1 as shown below, then we get: Y = A.1' + A'.1 = 0 + A' = A', which is a NOT gate or an inverter.

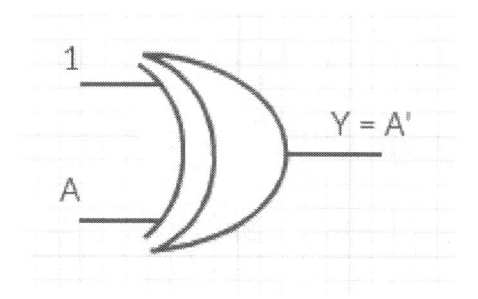

16. How can you design a pass gate or a buffer using XOR gate?

A pass gate or a buffer passes the input as it is to the output. If A is the input and Y is the output, this can be represented by Y=A. Hence, this can be implemented using XOR gate by connecting one of the inputs to be always zero as below: Y = 0.A' + 0'.A = 0 + 1.A = A

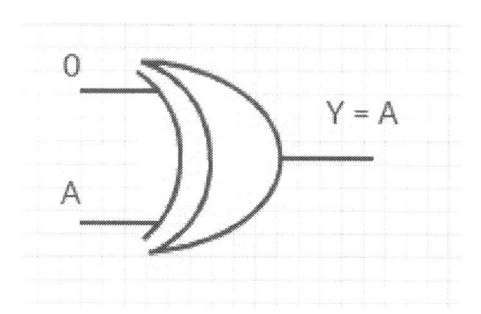

17. What will be the output of following gates if one of the inputs is unknown(x)?
   a) OR gate
   b) AND gate
   c) NOT gate
   d) XOR gate

a) If one of the inputs of OR gate is x, then the output depends on other input. If other input is one, the output of OR gate will be 1, and for any other values, output will be x.

b) If one of the inputs of AND gate is x, then the output depends on other input. If other input is zero, then the output of AND gate will be zero and for any other values, output will be x.
c) If the input of NOT gate is x, output will also be x.
d) If one of inputs of XOR gate is x, output will also be x.

18. A bulb in a staircase has two switches, one switch being at the ground floor and the other one at the first floor. The bulb can be turned ON and also can be turned OFF by any one of the switches irrespective of the state of the other switch. Which gate does this logic resemble for the bulb turning on?

Let us take S0 and S1 as the two switches. If already a switch is off (0), then changing other switch to 1 should give ON=1. If already a switch is ON (1), then changing other switch to 1 should turn off the bulb (OFF=1).

Accordingly, you can have following table representing how the ON/OFF behaves based on switch.

S0	S1	ON	OFF	
0	0	0	1	
0	1	1	0	(S0 was off, S1=1 causes bulb to turn on)
1	0	1	0	(S1 was off, S0=1 causes bulb to turn on)
1	1	0	1	(either S0/S1 was on and changing other switch to 1 causes bulb to be off)

Hence, the turning on of bulb behaves like an XOR gate.

1.3 Combinational Logic Circuits

19. What is the difference between Combinational and Sequential circuits?

A circuit whose output at any instant depends only on the inputs at the present instant of time is called a combinational circuit. Hence, these circuits do not contain any memory elements. Some examples of combinational circuits are Half Adder, Full Adder, Multiplexer, Decoder etc.

A circuit whose output at any instant depends both on the inputs at the present instant of time as well as output values from the past is called a sequential circuit. These circuits hence have some form of memory elements to remember the past values. Some examples of sequential circuit are Flip-flops, Registers, Counters, etc.

20. What is the difference between a Multiplexer and Demultiplexer?

An **n to 1** multiplexer, or MUX, for short, is a device that allows you to pick one of **n** inputs and direct it to a single output.

Demultiplexers (or DeMUX for short) are basically multiplexers where the inputs and outputs are reversed. For a **1 to n** DeMUX, you have a single input, and **n** outputs to choose for directing the input.

21. Design a 4:1 MUX using 2:1 multiplexers?

A 4:1 MUX can be designed using 2:1 MUX as shown below. Following is the truth table for a 4:1 MUX (S1, S0 are the select lines and I0-I3 are the 4 input lines while Y is the output).

S1	S0	Y
0	0	I0
0	1	I1
1	0	I2
1	1	I3

So you can see that S1 can be used to select one half of the inputs and within each half S0 can be used to select one of the two inputs within that half.

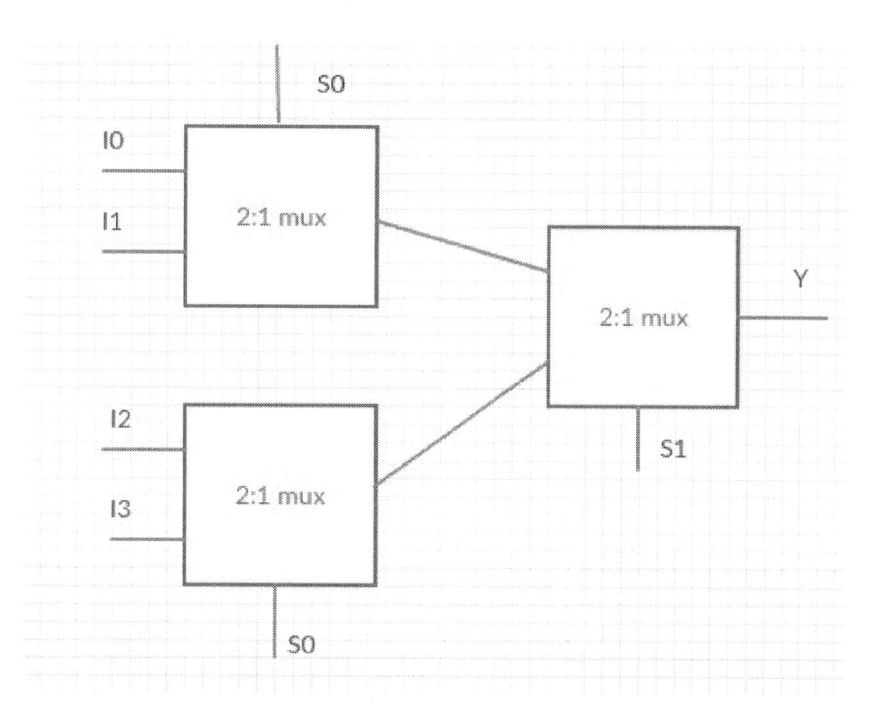

22. How many 2:1 multiplexers will you need to design a 2^n:1 MUX?

As we saw in previous question, a 4:1 MUX can be implemented using three 2:1 MUXes. This includes one MUX to select the two halves and two MUXes to select between two inputs of each half of inputs.

If we extend this concept to a 8:1 MUX, we will need one 2:1 MUX to select between two halves (4 inputs each) and then we will need two 4:1 MUXes (each requiring three 2:1 MUXes).

Hence, for a "2^n:1" MUX, we will require one 2:1 MUX and two "n:1" MUXes. And if we apply the same logic for n:1 MUX and further, we will get the following equation which is a geometric progression.

Total number of MUXes needed = $1 + 2 + 4 + 8 + \ldots + 2^{(n-1)} = \mathbf{2^n - 1}$

23. What is the difference between an encoder and a decoder?

A decoder is a combinational circuit that decodes a given number of inputs into a given number of output signals.

For example: A 3 to 8 decoder will decode a 3 bit input signal to an 8 bit output signal as follows

000 => Out0
001 => Out1
010 => Out2
011 => Out3
100 => Out4
101 => Out5
110 => Out6
111 => Out7

And encoder is a combinational circuit that does the other way around. It takes a given number of **n** inputs and encodes them into a smaller number of outputs

For example: An 8 to 3 encoder could do exactly reverse of above 3 to 8 decoder. There can be 8 inputs and each of them can be encoded into a 3 bit binary output.

24. How is an encoder different from a multiplexer?

An encoder is a similar to a multiplexer with the difference that a multiplexer only has a single output to which n inputs are multiplexed while an encoder normally has 2^n inputs (or less) and n outputs.

25. What is a priority encoder and how is it different from a simple encoder?

A simple encoder is a circuit that converts a 2^n bit one-hot vector to an n-bit output.
For example: a 4 to 2 simple encoder encodes as per following table. The simple encoder expects the inputs to be one hot and if more than one input is high, then the outputs becomes X.

4 to 2 Simple Encoder

I_3	I_2	I_1	I_0	O_1	O_0	V
0	0	0	0	X	X	0
0	0	0	1	0	0	1
0	0	1	0	0	1	1
0	1	0	0	1	0	1
1	0	0	0	1	1	1

A priority encoder on the other hand encodes inputs with more than one bit being high using a priority. For example: a 4 to 2 priority encoder will encode a 4 bit input and if more than one bit is high, the MSB takes priority. Hence, it can be represented as table shown below.

4 to 2 Priority Encoder

I_3	I_2	I_1	I_0	O_1	O_0	V
0	0	0	0	X	X	0
0	0	0	1	0	0	1
0	0	1	X	0	1	1
0	1	X	X	1	0	1
1	X	X	X	1	1	1

26. **What is a ring oscillator? What would be the frequency of a ring oscillator implemented using three NOT gates if each gate has a delay of 2 ps?**

A ring oscillator is a device composed of an **odd** number of NOT gates whose output oscillates between two voltage levels, representing true and false. The NOT gates, or inverters, are attached in a chain and the output of the last inverter is fed back into the first.

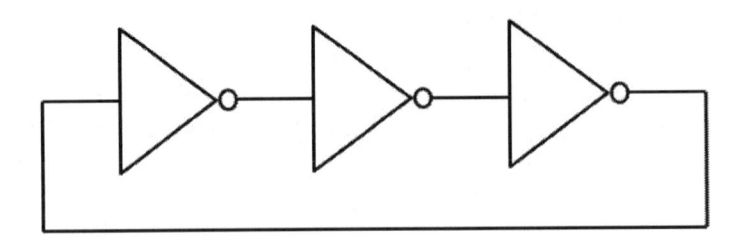

If three NOT gates are connected in a chain, then it would take three times the inverter delay for a value from input to transition to the output. Therefore, for two transitions it takes 6 times inverter delay. Hence the clock frequency will be 1/ [6*(inverter delay)]. For our present case, clock frequency will be = 1/(6*2) THz = 1000/12 GHz = 83.33 GHz.

1.4 Sequential Circuits and State Machines

27. What is the difference between Synchronous and Asynchronous circuits?

Sequential circuits can be of two types - **Synchronous** circuits and **Asynchronous** circuits. Synchronous sequential circuits change their states and output values at discrete instants of time, which are specified by the rising (transition from 0 to 1) or falling edge (transition from 1 to 0) of a clock signal. A simple example is a flip-flop which stores a binary value and can change on an edge of clock based on input values.

In Asynchronous sequential circuits, the transition from one state to another is initiated by the change in the primary inputs without any external synchronization like a clock edge. It can be considered as combinational circuits with feedback loop. Because of the feedback among logic gates, asynchronous sequential circuits may, at times, become unstable due to transient conditions and are not used commonly. A simple example: RS Latch.

28. Explain the concept of "Setup" and "Hold" times?

Setup time is the minimum amount of time during which data signal should be stable before the clock makes a valid transition. **Hold time** is the minimum amount of time during which data signal should be stable after the clock makes a valid transition.

29. What is meant by clock skew?

The difference in arrival times of the clock signal at any two flops which are interacting with one another is referred to as clock skew

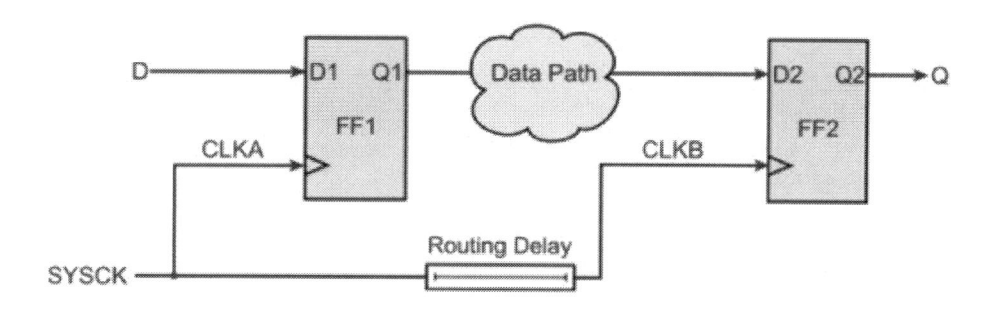

For example in the above diagram, the D input from first flip-flop propagates through a combinational datapath circuit to second flip-flop. A clock from a common source (SYSCK) is routed to both flip-flops, but because of wire or routing delay, there could be a small difference when the edges are seen on the two flip-flops. The difference of this time is known as clock skew.

The clock skew is only important between two flip-flops where one flip-flop output is being sampled by the second flip-flop.

30. **For a given sequential circuit as shown below, assume that both the flip flops have a clock to output delay = 10ns, setup time=5ns and hold time=2ns. Also assume that the combinatorial data path has a delay of 10ns. Calculate the maximum frequency of CLKA that is possible for design to operate correctly?**

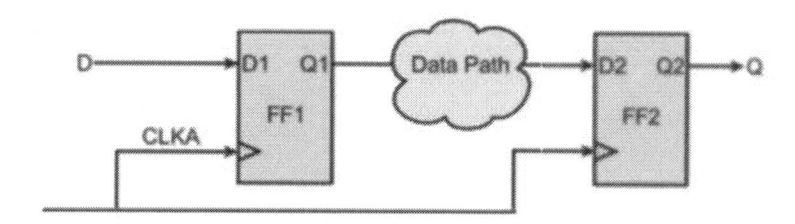

For this sequential circuit to operate correctly, output of the first flip-flop should propagate through the combinatorial data path and should be stable for a minimum duration equal to the setup time of the second flip-flop before the next clock edge.

If T_{CLKA} is the clock period, T_{CQ} is the clock to output delay, T_{PD} is the propagation delay for the data path and T_{SET} is the set up time of flip flop, then we have this condition as $T_{CLKA} \geq T_{CQ} + T_{PD} + T_{SET}$

Hence, the clock period in this example needs to be >= 10+10+5 = 25ns and the max frequency will be 1/25ns = 40MHz.

31. What is the difference between a flip-flop and a latch?

Latches and flip-flops are the basic elements for storing information. One latch or flip-flop can store one bit of information.
The main difference between latches and flip-flops is that for latches, their outputs are constantly affected by their inputs as long as the enable signal is asserted. In other words, when the enable signal is high, the contents of latches changes immediately when inputs changes.
Flip-flops, on the other hand, will change the contents only at the rising or falling edge of the enable signal which is usually a clock signal. After the rising or falling edge of the clock, the flip-flop content remains constant even if the input changes.

32. What is a race condition? Where does it occur and how can it be avoided?

When an output has an unexpected dependency on relative ordering or timing of different events, a race condition occurs. With respect to Digital Electronics and SystemVerilog, race conditions can be classified into two types:
  1) Hardware race condition
  2) Simulation Induced race condition.

Hardware race condition: If we look at SR latch (NAND gate based) below, if both the inputs are 1 (S=1 and R=1), Q and Q' both would become 1 and then feedback into NAND gates. Now, if both S and R are immediately changed to 0, Q and Q' both would enter into a race condition (value would start oscillating: 1 to 0, 0 to 1, 1 to 0, and so on...).

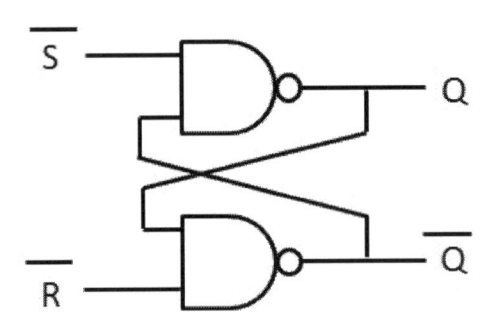

Hardware race condition can be avoided by proper design techniques. For Example: In this case of SR latch, we can avoid race condition by having an enable/control signal as shown below:

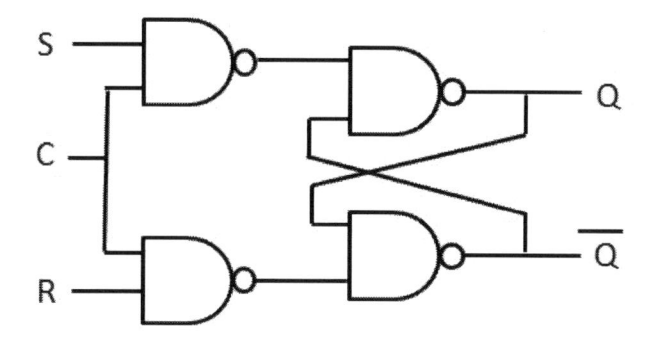

Simulation induced race condition: Look at the SystemVerilog code below:
```
always @(posedge clk or posedge reset)
  if (reset)  X1 = 0; // reset
  else X1 = X2;
always @(posedge clk or posedge reset)
  if (reset) X2 = 1; // reset
  else X2 = X1;
```

SystemVerilog simulators don't guarantee any execution order between multiple always blocks. In above example, since we are using blocking assignments, there can be a race condition and we can see different values of X1 and X2 in multiple different simulations. This is a typical example of what a race condition is. If the first always block gets executed before second always block, we will see X1 and X2 to be 1. If the second always block gets executed before first always block, we will see both X1 and X2 to be zero.
There are many coding guidelines following which we can avoid simulation induced race conditions. For Example: This particular race condition can be avoided by using nonblocking assignments instead of blocking assignments.

33. Implement D flip-flop using 2:1 MUX.

Let us take the case of negative edge triggered D Flip-flop. A negative edge triggered D Flip-flop transfers the input (D) to the output at a negative clock edge only. At all the other times, there is no change in the output (Q retains its value). To implement this functionality using 2:1 MUX, we need two 2:1 MUXes as shown in the figure on the next page.

Here, one MUX has select line as CLK signal and the other MUX has select line as CLK' signal (inverted CLK). Now, when CLK is 1, select line of the first MUX (one on the LHS) selects D (connected to i1), and D is thus transferred to the output of first MUX (this output is further is fed-back to the first MUX and is also connected to i1 of second MUX). Since, CLK' is 0 when CLK is 1, D is not transferred to Q (as i0 is selected for second MUX). Hence, till the time CLK stays 1, any change in D is reflected at the output of first MUX but is not reflected at the output of second MUX as i0 is selected for second MUX (which is nothing but Q as Q is connected to i0 for second MUX). Now, when the CLK transition from 1 to 0, D is no longer selected (i0 is active for first MUX) and the value of D which was present just before the negative edge of the clock is transitioned to the output Q.

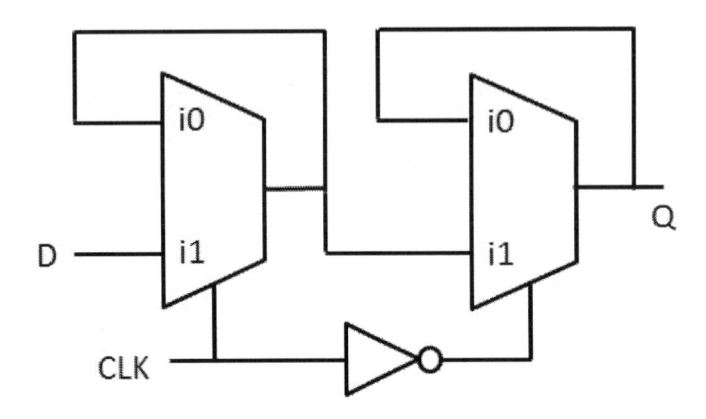

34. How can you convert a D Flip-flop to a T Flip-flop?

For converting any flip-flop (say D Flip-flop) to any other flip-flop (say T Flip-flop), write down the excitation table of the target flip-flop behavior (T flip-flop in this case) and implement a combinational logic which does the conversion of inputs of T Flip-flop to correct inputs for D Flip-flop such that together it behaves as a T flip-flop. Following diagram illustrates this.

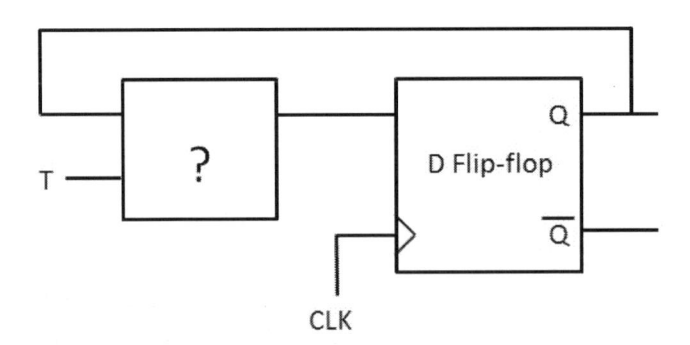

Following table consists of excitation table of T Flip-flop (first 3 columns below) and the last column shows the values of D that is needed for the outputs Qn to change to Qn+1 .

T	Qn	Qn+1	D
0	0	0	0
1	0	1	1
1	1	0	0
0	1	1	1

Looking at above table, we implement D as a function of T, Q(n), and Q(n+1) using K-maps. This would give us: **D = T.Qn' + T'.Qn, which is an XOR gate.**

35. Convert a JK flip-flop to D Flip-flop.

Following the principle described in the above question, we identify the combinational logic that is needed for conversion. Answer would be: **J = D and K = D'**

36. What is difference between a synchronous counter and an asynchronous counter?

A counter is a sequential circuit that counts in a cyclic sequence which can be either counting up or counting down. Counters are implemented using a number of flip flops and combinational logic that feeds output of one flip-flop to another. There are two types of counters - synchronous and asynchronous.
In synchronous counters, the clock inputs of all flip-flops are connected to a common clock signal and hence all flip-flops changes synchronously.
In asynchronous counters, the clock input is connected only to the first flip-flop and the output for first flip-flop is connected to the clock input of second flip-flop and similarly every other flip-flop is clocked by the output of previous flip-flop.
Some examples of synchronous counters are Ring counter and Johnson counter while some examples for asynchronous counters are Up-Down counters.

37. What is the difference between "Ripple Carry adder" and "Carry Look-ahead adder"?

Ripple carry adders are slow adders because the inputs ($A_1....A_N$, $B_1....B_N$) and carry-in's ($C_1....C_N$) ripple "leftwards" until the final Carry-Out (C_{OUT}) and most significant bit of the Sum (i.e. S_N) are generated. This is because each carry bit is calculated along with the sum bit and each bit must wait until the previous carry has been calculated in order to begin calculation of its own sum bit and carry bit. This is represented by diagram shown below.
Carry Look-ahead adders are fast adders as they reduce the time required to calculate carry bits. It calculates carry bits before the sum bits and this reduces wait time for calculating other significant bits of the sum. To facilitate this operation, intermediate "Propagate" and "Generate" functions are used, where Generate (G_i) = A_iB_i and Propagate (P_i) = $A_i \wedge B_i$
Generate function G_i produces 1 when there is a carry out from position i, and propagate function P_i is true when incoming carry is propagated. Hence, $C_{i+1} = G_i + P_iC_i$
This is represented by diagram shown below.

<u>N-Bit Ripple Carry Adder</u>

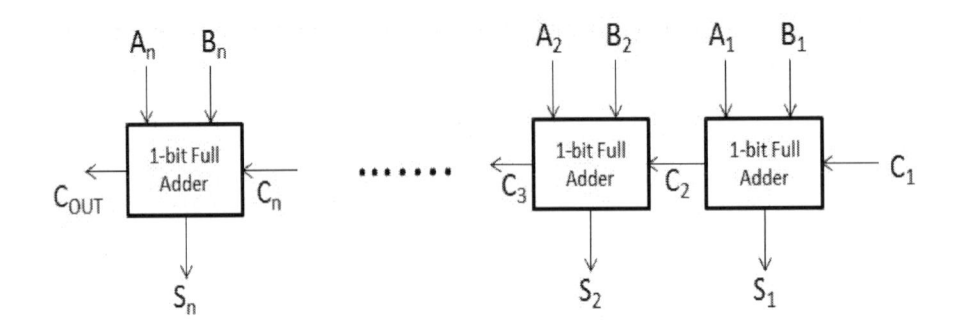

N-Bit Carry Look-Ahead Adder

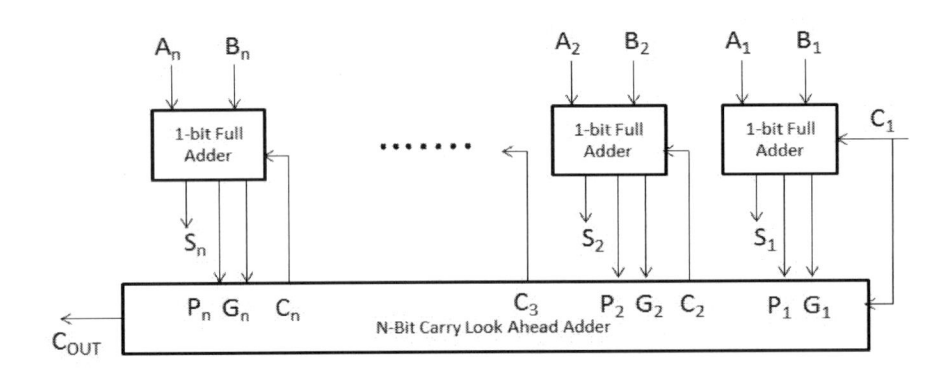

38. What is the longest path for N-bit ripple carry adder?

(2N+1) Gates.
To understand this better, think in terms of 1-bit and 2-bit ripple carry adders.
For 1-bit ripple carry adder, longest path has 3 gates.
For 2-bit ripple carry adder, longest path has 5 gates.

39. What is the difference between synchronous and asynchronous reset?

A Reset is synchronous when it is sampled on a clock edge. When reset is synchronous, it is treated just like any other input signal which is also sampled on clock edge.
A reset is asynchronous when reset can happen even without clock. The reset gets the highest priority and can happen any time.

40. How many flip-flops are needed to implement a 32 bit register?

Each flip-flop can save one bit of information. Hence to implement a 32 bit register, we would need 32 flip-flops.

41. What is the difference between a Mealy and a Moore finite state machine?

A Mealy Machine is a finite state machine whose output depends on the present state as well as the present input.
A Moore Machine is a finite state machine whose output depends only on the present state.

42. What is an Excitation table?

An excitation table shows the minimum inputs necessary to generate the next state when the present state is known. For Example: below are Excitation Tables for D, T and JK Flip-flops. The current state is identified by Q(t) while next state is Q(t+1).

Q(t)	Q(t+1)	D	Operation
0	0	0	Reset
0	1	1	Set
1	0	0	Reset
1	1	1	Set

Q(t)	Q(t+1)	T	Operation
0	0	0	No Change
0	1	1	Complement
1	0	1	Complement
1	1	0	No Change

Q(t)	Q(t+1)	J	K	Operation
0	0	0	X	No Change/Reset
0	1	1	X	Set/Complement
1	0	X	1	Reset/Complement
1	1	X	0	No Change/Set

43. If given a choice, which flip-flop would you use to implement a synchronous circuit (say a sequence detector)? D or JK?

Depends on the usage scenario.
JK flip-flops can lead to a simpler circuit because there are many don't care values for the flip-flop inputs to achieve next state from a present state.
D flip-flops have the advantage that we don't have to setup flip-flop inputs at all as the next state for D flip-flop is equal to input.

44. In practice, which flip-flop is used more often for implementing a synchronous circuit? D or JK?

In practice, D flip-flops are used more often because:
1) There are no excitation tables to worry about (Next State = Input)
2) There is only one input for each flip-flop (not two as compared to JK)
3) D flip-flops are simpler to implement as compared to JK.

45. Design a sequence detector state machine that detects a pattern "10110" from an input serial stream. Use D Flip-Flops.

To design any basic sequential circuit, we need to perform following five steps:
1) Construct a state diagram and a state table,
2) Assign binary codes to all the states defined,
3) Use Present States, Next States, and flip-flop excitation table to find out the correct flip-flop input values that can help achieve the Next State from the Present State and Inputs,
4) Write equations for flip-flop inputs and outputs (simplify equations using circuit minimization techniques like K-Maps),
5) Build the circuit.

Let us assume following states and corresponding meanings:
A: None of the desired pattern is detected yet
B: First bit (1) of the desired pattern is seen
C: First two bits (10) of the desired pattern are seen
D: First three bits (101) of the desired pattern are seen
E: First four bits (1011) of the desired pattern are seen

Based upon the pattern and the states, following will be the state diagram:

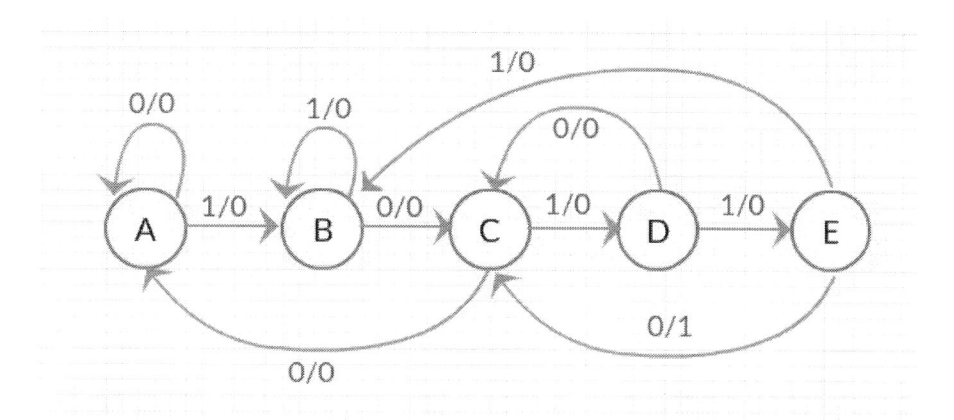

The tricky part of this state machine to understand is how it can detect start of a new pattern from the middle of a detection pattern. For example, notice the state transition from state D to C. This is required to detect the given pattern from a stream like- 1010110 - where a new match can start while state machine is D. Also, notice the state transition from state E to C which is needed to detect another continuing pattern once the match is seen on first 10110.

Since we have five states, we will need 3 D Flip-Flops to implement this sequence detector. Let inputs of D Flip-flops be D2, D1, D0. Further, let's assume input as X and output as Z.

Let state A be represented as 000, B as 001, C as 010, D as 011 and E as 100.

Now, using Present States, Next States and flip-flop excitation table, try to find that inputs of flip-flops that would result in transition from present state to next state. Minimize the circuit using K-Map minimization and get following:

Output (Z) = Q2.Q1'.Q0'.X'
D2 = Q2'.Q1.Q0.X
D1 = Q2.Q1'.Q0'.X' + Q2'.Q1.Q0'.X + Q2'.Q0.X'
D0 = Q2'.Q1'.X + Q2'.Q0'.X + Q1'.Q0'.X

46. What is the minimum number of flip-flops required to implement a digital synchronous circuit with 9 states?

Each flip-flop can store one bit of information. Hence one flip-flop can be used to implement a synchronous circuit requiring up-to 2 states, two flips-flops for synchronous circuits requiring up-to 4 states, three flips-flops for synchronous circuits requiring up-to 8 states, and four flip-flops for synchronous circuits requiring up-to 16 states. Hence, for a synchronous circuit with 9 states, we have to use 4 flip-flops.

47. Design a circuit that would count 1 every time another counter counts from 0 to 255.

Implement f/256 circuit.

48. Implement following logics using minimum number of D Flip-Flops:
   a) Clock Divide by 2
   b) Clock Divide by 4

Answer:

   a) Clock Divide by 2.

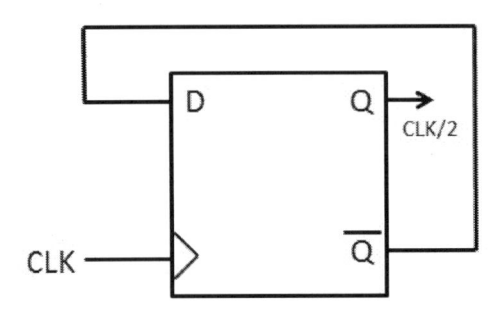

   b) Clock Divide by 4.

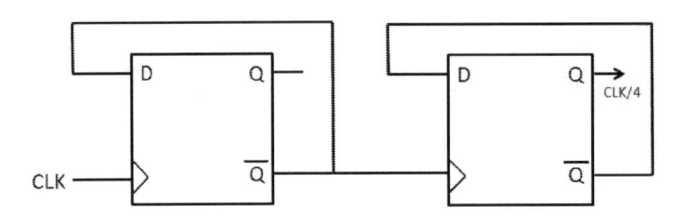

1.5 Other Miscellaneous Digital Design Questions

49. What is the difference between an ASIC and SOC?

An Application-Specific Integrated Circuit (ASIC) is a component that is designed for a specific application and is used by specific companies in a specific system or sold to a single company for their specific use. For Example: A specific 24x24 10G switch designed for a very specific system

An Application-Specific Standard Product (ASSP) is a more general-purpose device that is created using ASIC tools and technologies, but is intended for use by multiple companies for different systems in a wider market. For Example: An audio/video encoder/decoder chip which is also for a specific application but targets a wider market.

A System-on-Chip (SoC) is a chip that integrates an entire subsystem including a microprocessor or microcontroller, memory, peripherals, custom logic, and so forth. An ASIC or ASSP can either be an SOC or a non-SOC based on the different elements of the design present in the chip.

50. What are the different steps in a typical ASIC or SOC design flow?

The following block diagram describes a typical design flow for an ASIC or FPGA or SOC design.

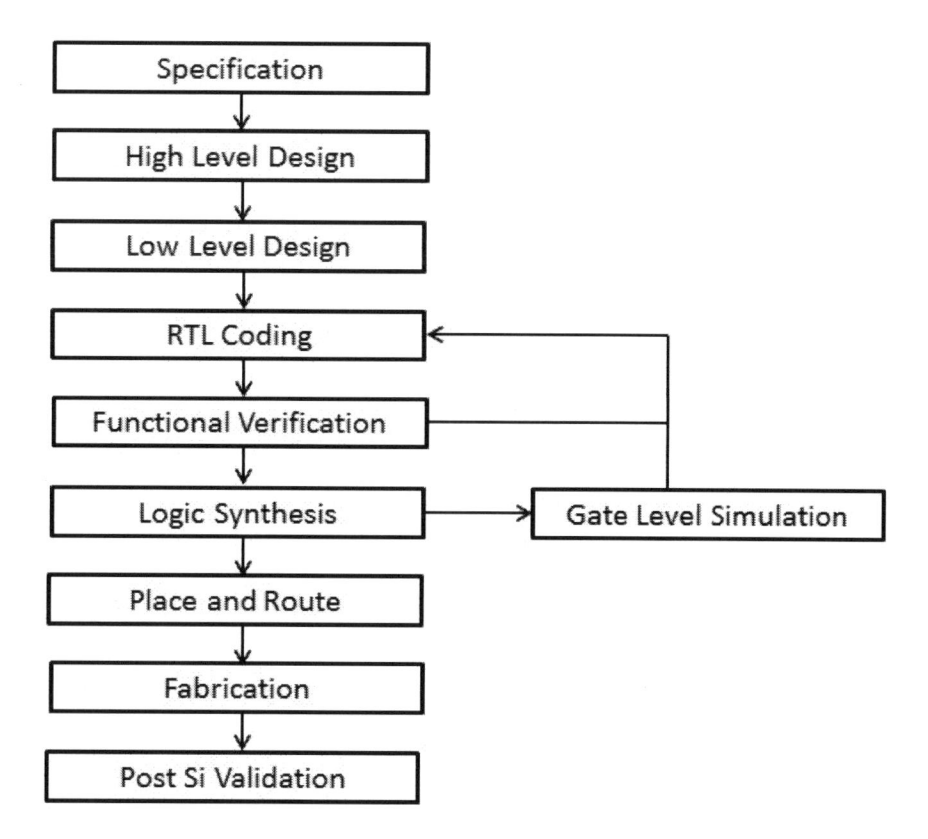

```
Specification
     ↓
High Level Design
     ↓
Low Level Design
     ↓
RTL Coding ←─────────────────┐
     ↓                       │
Functional Verification ─────┤
     ↓                       │
Logic Synthesis ──→ Gate Level Simulation
     ↓
Place and Route
     ↓
Fabrication
     ↓
Post Si Validation
```

1) **Specification:** This is the first stage in the design process where we define the important parameters of the system that has to be designed into a specification.
2) **High level Design:** In this stage, various details of the design architecture are defined. In this stage, details about the different functional blocks and the interface communication protocols between them etc. are defined.
3) **Low level Design:** This phase is also known as microarchitecture phase. In this phase lower level design details about each functional block implementation are designed. This can include details like modules, state machines, counters, MUXes, decoders, internal registers etc.
4) **RTL coding:** In RTL coding phase, the micro design is modelled in a Hardware Description Language like Verilog/VHDL, using synthesizable constructs of the

language. Synthesizable constructs are used so that the RTL model can be input to a synthesis tool to map the design to actual gate level implementation later.
5) **Functional Verification:** Functional Verification is the process of verifying the functional characteristics of the design by generating different input stimulus and checking for correct behavior of the design implementation.
6) **Logic Synthesis:** Synthesis is the process in which a synthesis tool like design compiler takes in the RTL, target technology, and constraints as inputs and maps the RTL to target technology primitives. Functional equivalence checks are also done after synthesis to check for equivalence between the input RTL model and the output gate level model.
7) **Placement and Routing**: Gate-level netlist from the synthesis tool is taken and imported into place and route tool in the Verilog netlist format. All the gates and flip-flops are placed, Clock tree synthesis and reset is routed. After this each block is routed, output of the P&R tool is a GDS file, which is used by a foundry for fabricating the ASIC.
8) **Gate level Simulation:** The Placement and Routing tool generates an SDF (Standard Delay File) that contains timing information of the gates. This is back annotated along with gate level netlist and some functional patterns are run to verify the design functionality. A static timing analysis tool like Prime time can also be used for performing static timing analysis checks.
9) **Fabrication**: Once the gate level simulations verify the functional correctness of the gate level design after the Placement and Routing phase, then the design is ready for manufacturing. The final GDS file (a binary database file format which is the default industry standard for data exchange of integrated circuit or IC layout artwork) is normally send to a foundry which fabricates the silicon. Once fabricated, proper packaging is done and the chip is made ready for testing.
10) **Post silicon Validation:** Once the chip is back from fabrication, it needs to be put in a real test environment and tested before it can be used widely in the market. This phase involves testing in lab using real hardware boards and software/firmware that programs the chip. Since the speed of simulation with RTL is very slow compared to testing in lab with real silicon, there is always a possibility to find a bug in silicon validation and hence this is very important before qualifying the design for a market.

Chapter 2: Computer Architecture

Thanks to Moore's law and continuous innovations in semiconductor technology, digital VLSI system designs are integrating more and more components to a single chip. Digital IC designs are trending to be more and more SOC (System on Chip) designs that integrate a microprocessor or a Micro controller along with other processors like GPU or DSP and numerous system components. System components could be hardware accelerators, memory controllers, peripherals and controllers like PCIE, USB, SATA, Ethernet etc. Another trend is the increasing number of processor cores (from dual to quad to octa cores) integrated on a single chip.

With this trend, understanding fundamentals of computer architecture has gained a lot of importance for VLSI design and verification engineers. Most of the SOC design verification revolves around a CPU and may involve writing tests that can program or initialize CPUs and access other system components using the CPU. Hence, it is observed that in many SOC design or verification interviews, candidates are judged based on their computer architecture knowledge.
In this section, we list down some of the most commonly asked questions in computer architecture.

51. What is the difference between a RISC and CISC architecture?

RISC refers to Reduced Instruction Set Computing and CISC refers to Complex Instruction Set Computing.

- RISC architecture has less number of instructions and these instructions are simple instructions (i.e. fixed length instructions, and less addressing modes). On the other hand, CISC architecture has more number of instructions and these instructions are complex in nature (i.e. variable length instructions, and more addressing modes).
- RISC approach is to have smaller instructions and less complex hardware, whereas CISC approach is to have more complex hardware to decode and break down complex instructions. Hence, In RISC architecture emphasis is more on software, whereas in CISC architecture emphasis is more on hardware.
- Since CISC has complex hardware, it requires smaller software codes and hence less RAM to store programming instructions. As RISC has less complex hardware, RISC require software programs that uses more number of instructions and hence more RAM to store instructions.
- Instructions in RISC architecture usually require one clock cycle to finish, whereas instructions in CISC architecture can take multiple clock cycles to complete depending upon the complexity of the instruction. Due to this, pipelining is possible in RISC architecture.
- RISC architecture aims to improve performance by reducing number of cycles per instruction whereas CISC architecture attempts to improve performance by minimizing number of instructions per program.

CISC architectures supports single instruction that can read from memory, do some operation and store back to memory (known as memory to memory operation).
RISC architectures on the other hand would need multiple instructions to: 1) load the value from memory to an internal register, 2) perform the intended operation, and 3) write the register results back to memory.

Example: If we have to multiply two numbers stored at memory locations M1 and M2 and store the result back in memory location M1, we can achieve this through a single CISC instruction:
```
MULT M1, M2
```
Whereas for RISC, we would need following multiple instructions:
```
LOAD A, M1
LOAD B, M2
PROD A, B
STORE M1, A
```

Having mentioned all the differences above, it's important to point out that with advanced computer micro-architecture; even lots of CISC architecture implementations internally translate the complex instructions into simpler instructions first.

52. What is the difference between Von-Neumann and Harvard Architecture and which would you prefer?

In **Von Neumann architecture**, there is a single memory that can hold both data and instructions. Typically, this would mean that there is a single bus from CPU to memory that accesses both data and instructions. This architecture has a unified cache for both data and instructions.
In **Harvard Architecture**, memory is separate for data and instructions. There can be two separate buses to access data and instruction memory simultaneously. There will also be separate caches for Instruction and Data in this architecture.
The Von Neumann architecture is relatively older and most of the modern computer architectures are based on Harvard architecture.

53. Explain the concept of Little Endian and Big Endian formats in terms of memory storage?

Endian-ness refers to the order in which bytes are stored in a memory (It can also be applicable to digital transmission systems where it describes the byte order for transmission)

Memory is normally byte addressable but majority of the computer architectures works on 32 bit size or a word size (4 bytes) operands. Hence, for storing a word into a byte addressable memory there are two ways:
1) Store the Most significant byte of the word at a smaller address. This type of storage refers to Big Endian format.
2) Store the Least significant byte of the word at a smaller address. This type of storage refers to Little Endian format.

For example: If a CPU is trying to write the word 0xDDCCBBAA to an address starting from 0x1000 (address range: 0x1000 to 0x1003), the bytes can be stored in following two different endianness as shown below.

Little Endian	
Address	Data Byte
0x1000	AA
0x1001	BB
0x1002	CC
0x1003	DD

Big Endian	
Address	Data Byte
0x1000	DD
0x1001	CC
0x1002	BB
0x1003	AA

54. What is the difference between a SRAM and a DRAM?

DRAM stands for Dynamic Random Access Memory. It is a type of memory in which the data is stored in the form of a charge. Each memory cell in a DRAM is made of a transistor and a capacitor. The data is stored in the capacitor. DRAMs are volatile devices because the capacitor can lose charge due to leakage. Hence, to keep the data in the memory, the device must be regularly refreshed.

On the other hand, SRAM is a static memory and retains a value as long as power is supplied. SRAM is typically faster than DRAM since it doesn't have refresh cycles. Each SRAM memory cell is comprised of 6 Transistors (unlike a DRAM memory cell which is comprised of 1 Transistor and 1 Capacitor). Due to this, the cost per memory cell is more for SRAM

In terms of usage, SRAMs are used in Caches because of higher speed and DRAMs are used for main memory in a PC because of higher densities.

55. A computer has a memory of 256 Kilobytes. How many address bits are needed if each byte location needs to be addressed?

Since the total memory size is 256KB ($2^8 * 2^{10}$ Bytes), each address would be **eighteen** bits wide.

56. What are the different types of registers implemented in a CPU?

1) Program Counter (PC): A Program Counter is a register that holds the address of the instruction being executed currently.
2) Instruction Register (IR): An Instruction Register is a register that holds the instruction that is currently getting executed. (It will be value at the address pointed by PC)
3) Accumulator: An accumulator is a register that holds the intermediate results of arithmetic and logic operations inside a processor

4) **General Purpose Registers**: General Purpose Registers are registers that can store any transient data required by a program. The number of General purpose registers is defined by the architecture and these can be used by Software (Assembler) for storing temporary data during a program execution. More the number of General Purpose registers, faster will the CPU execution.
5) **Stack Pointer Register (SP)**: The Stack Pointer Register is a special purpose register that stores the address of the most recent entry that was pushed on to stack. The most typical use of a stack is to store the return address of a subroutine call. The SP register helps in maintaining the top of the Stack Address.

57. Explain the concept of pipelining in computer architecture?

Pipelining is a technique that implements a form of parallelism called instruction-level parallelism within a single processor. The basic instruction cycle is broken up into a series of steps called a pipeline. Rather than processing each instruction sequentially (finishing one instruction before starting the next), each instruction is split up into a sequence of steps so different steps can be executed in parallel and instructions can be processed concurrently (starting one instruction before finishing the previous one).

Pipelining increases instruction throughput by performing multiple operations at the same time, but does not reduce instruction latency, which is the time to complete a single instruction from start to finish, as it still must go through all steps.

For example: an Instruction life cycle can be divided into five stages - Fetch, Decode, Execute, Memory access, and Write back. This allows the processor to work on several instructions in parallel.

58. What is a pipeline hazard? What are the different types of hazards in a pipelined microprocessor design?

A pipeline hazard is a situation where the next instruction in a program cannot be executed for a certain reason. There are three types of hazards that occur in a pipelined microprocessor as follows:

1) **Structural Hazards**: These hazards arise because of resource conflict that prevents overlapped execution. For example: if the design has a single Floating Point Execution unit and if each execution takes 2 clock cycles, then having back to back Floating point instructions in the program will cause the pipeline to stall. Another resource that can conflict is memory/cache access.
2) **Data Hazards**: These hazards arise when an instruction depends on the result of a previous instruction in a way exposed by the pipeline overlapped execution. There can be three types of Data hazards:
    a) Read after Write (RAW) - This happens if an instruction needs a source which is written by a previous instruction.
    b) Write after Write (WAW) - This happens if an instruction writes to a register which is also written by a previous instruction
    c) Write after Read (WAR) - This happens if an instruction writes to a register

which is a source for a previous instruction
3) **Control Hazards:** These hazards arise because of branch and jump instructions that changes the sequence of program execution.

59. What techniques can be used to avoid each of the 3 types of pipeline hazards - Structural, Data and Control Hazards?

Following are some of the techniques that are used to avoid each of the pipeline hazards:
1) **Structural Hazards**:
    a) Duplicating resources to enable parallel execution - separating instruction and data caches, having multiple execution units for integer and floating point operations, separate load and store units, etc.
2) **Data Hazards**:
    a) Out of order execution - Instructions which are not dependent on each other can execute while the dependent instruction stalls.
    b) Data forwarding - For RAW hazards, the write from an instruction can be forwarded to the next dependent instruction to eliminate the hazard.
3) **Control Hazards**:
    a) Use branch prediction algorithms to make predictions about branch outcome so that correct set of instructions can be fetched following the branch.

60. A pipelined machine has 10 stages as shown below. Each stage takes 1 ns to process a data element. Assuming there are no hazards, calculate the time taken to process 100 data elements by the machine.

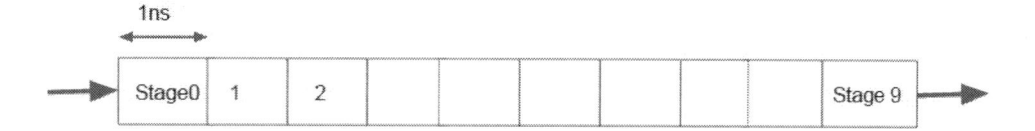

```
1ns
<---->
  Stage0 | 1 | 2 |   |   |   |   |   |   | Stage 9
```

Each stage of pipeline takes 1ns to process a data element. Since there are 10 stages, the first element takes 10 * 1ns to come out of the pipeline and by that time, the pipeline would be full and all other 99 elements would only take 1ns each. Hence total time taken = (10+99) ns = **109 ns**

61. What are the different types of addressing modes for an instruction?

Following are some of the most commonly used addressing modes for an instruction (though several other modes might also be supported by some architectures):
1) **Immediate mode**: In this mode, the operands are part of the instruction itself as constants as shown below:
    add r0 r1 0x12   (add a constant value of 0x12 with contents of r1 and store result in r0)

2) **Direct Addressing mode:** In this mode, the address of the operand is directly specified in the instruction.
    ```
    load   r0    0x10000
    ```
    (load data from address 0x10000 to register r0)
3) **Register Addressing mode:** In this mode, the operands are placed in registers and the register names are directly specified part of instruction
    ```
    mul   r0, r1 , r2
    ```
    (multiply contents of r1 and r2, and store the result in r0)
4) **Indexed Addressing mode:** In this mode, content of an index register is added with an offset (which is part of instruction) to get the effective address.
    ```
    load   r0   r1   offset
    ```
    (Here r1 contains the base address, and "r1 + offset" will give the address of a memory location from which data is read and stored into r0)

62. What is the principle of spatial and temporal locality of reference?

Locality of reference is a principle that defines if a memory location is accessed by a program, how frequently will the same memory location or nearby storage locations be accessed again.
There are two types of locality of reference explained as below:
1) Temporal Locality: If at one point in time a particular memory location is referenced, then it is likely that the same location will be referenced again in the near future.
2) Spatial Locality: If a particular memory location is referenced at a particular time, then it is likely that nearby memory locations will be referenced in the near future.

63. What are different kinds of memories in a system?

1) Register
2) Cache
3) Main Memory/Primary Memory
4) Secondary Memory (Magnetic/Optical)

64. What is a cache?

Cache is a small amount of fast memory. It sits between the main memory and the CPU. It may also be located on a CPU chip/module.

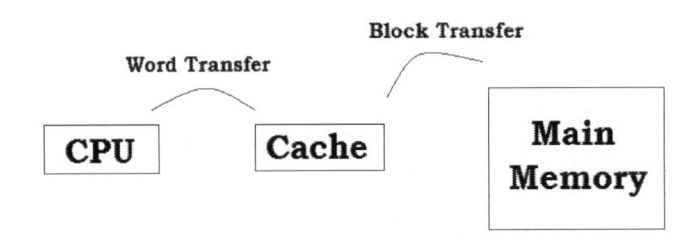

```
                              Block Transfer
         Word Transfer         ⌒
          ⌒
  ┌─────┐     ┌───────┐     ┌────────┐
  │ CPU │     │ Cache │     │  Main  │
  └─────┘     └───────┘     │ Memory │
                            └────────┘
```

65. Give an overview of Cache Operation. What's the principle behind functioning of cache?

Whenever CPU requests for contents of a memory location, cache is checked for this data first. If data is present in the cache, CPU gets the data directly from the cache. This is faster as CPU doesn't need to go to main memory for this data. If data is not present in the cache, a block of memory is read from main memory to the cache and is then delivered from cache to CPU in chunks of words. Cache includes tags to identify which block of main memory is in each cache slot.

66. What is a cache miss or hit condition?

When an address is looked up in the cache and if the cache contains that memory location, it is knows as a cache hit. When the address looked up in the cache is not found, then it is known as a cache miss condition.

67. Will there be a difference in the performance of a program which searches a value in a linked list vs a vector on a machine that has cache memory present?

A linked list is a data structure that stores its elements in non-contiguous memory location while a vector is a data structure that stores elements in contiguous locations.
For a design with cache memory: if one of the memory locations is present in the cache, it is highly likely that the following bytes (contiguous bytes) would also be present in cache memory as any fetch from the main memory to the cache is usually fetched in terms of cache lines (which are generally 64 or 128 bytes). Because of this, searching through a vector will be faster than searching through a linked list on a machine which has cache memory.

68. What are the different mechanisms used for mapping memory to caches? Compare the advantages and disadvantages of each.

There are three main mapping techniques that are used for mapping main memory to caches as explained below. In each of these mapping, the main memory and the cache memory are divided into blocks of memory (also known as cache line and is generally 64 bytes) which is the minimum size used for mapping.

1) **Direct Mapping**: In Direct mapping, there is always a one to one mapping between a block of main memory and cache memory. For example: in the diagram below, the size of cache memory is 128 blocks while the size of main memory is 4096 blocks. Block 0 of main memory will always map to Block 0 of cache memory, Block 1 to Block 1, ..., and Block 127 will map to Block 127. Further, Block 128 will again map to Block 0, Block 129 to Block 1, .., so on, and this can be generalized as Block " k" of main memory will map to Block "k modulo 128" on to the cache.

If the block size is 64B and the address is 32 bits, then address [5:0] will be used to index into the block, address [12:6] will be used to identify which block in cache this address can map and remaining address bits address [31:13] will be stored as tag bits along with the data in the cache memory.

Memory Address

Tag	Block Identifier	Block Index (Word Identifier In a Block)

Cache

Tag	Block 1
Tag	Block 2
Tag	Block 3
Tag	Block 4
...	...
Tag	Block 128

Main Memory

Block 1
Block 2
Block 3
Block 4
...
Block 4095

This is the simplest of all mapping and by knowing the memory address, the possible location in cache can be computed easily and a comparison with tag bits in that single location alone will tell you if a cache is hit or miss. The disadvantage of this mapping is that even though cache may not be full, but if memory access pattern is to addresses which fall in same block, it can cause more evictions and is not efficient.

2) **Fully Associative Mapping:** In fully associative mapping, any block of memory can be mapped to any block in the cache memory. Using the same example as shown in the diagram above, address [5:0] will be used to index inside the block, and all remaining bits i.e. address [31:6] will be stored as tag bits along with the data in cache. For looking up any memory address, all the address bits [31:6] have to be compared against all the tag bits in the cache location and this demands a bigger

comparator logic that can consume more power. The advantage of this mapping is that all locations can be fully utilized as any memory block can map to any cache block.

3) **Set-Way Associative Mapping:** In this mapping, the blocks of cache memory are grouped into a number of sets. For example, the diagram below shows the same cache of 128 blocks organized as 64 sets with each set having 2 blocks. Based on the number of blocks in a set, this is known as 2 way set associative cache. In this mapping, the main memory block is direct mapped to a set and then within the set it is associated with any block.

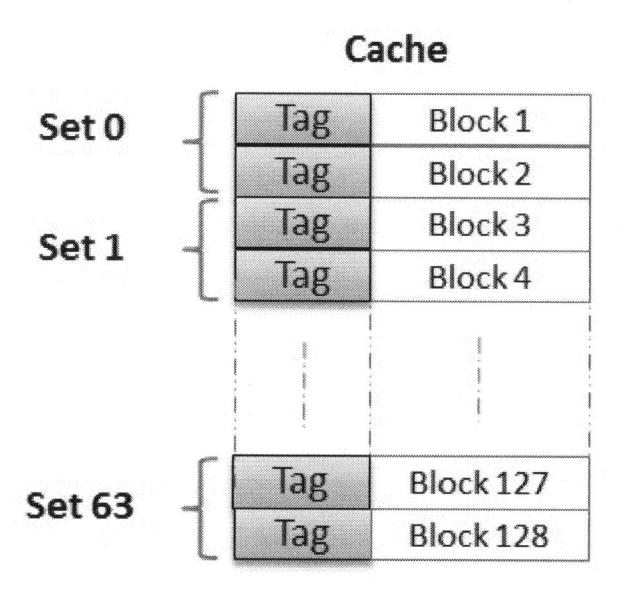

Cache

Set 0	Tag	Block 1
	Tag	Block 2
Set 1	Tag	Block 3
	Tag	Block 4
...	...	...
Set 63	Tag	Block 127
	Tag	Block 128

Considering the same example of a 32 bit address, address [5:0] will be used to index to a byte in the block, address [11:6] will be used to directly map to one of the 64 sets of the cache and remaining address bits [31:12] will stored as tag bits along with each cache line.

69. What's the disadvantage of having a cache with more associativity?

A cache with more associativity will need a bigger comparator to compare an incoming address against the tags stored in all the ways of the set. This means more power consumption and more hardware.

70. A byte addressable CPU with 16-bit address bus has a cache with the following characteristics: a) It is direct-mapped with each block of size 1 byte and b) The cache index for blocks is the four bits. How many blocks does the cache hold? How many tag bits are needed to be stored as part of cache block?

Since the index for blocks in cache is 4 bits, there will be a total of 16 blocks in the cache. Given a 16 bit address and block size of 1 byte, address [3:0] will be used to index into the 16 blocks in cache and remaining bits address[15:4] will be used as tag bits.

71. A 4-way set associative cache has a total size of 256KB. If each cache line is of size 64 bytes, how many sets will be present in the cache? How many address bits are needed as tag bits? Assume address size as 32 bits.

Total number of blocks in cache = 256K/64 = 4096. Since the cache is 4 way set associative, number of sets = 4096/4 = 1024 sets.
Given a 32 bit address and 64 byte cache line, address [5:0] is used to index into cache line, address [15:6] is used to find out which set the address maps to (10 bits) and remaining address bits [31:16] are used as tag bits.

72. What is difference between write-thru and write-back caches? What are the advantages and disadvantages?

Write Thru Cache: In a write-thru cache, every write operation to the cache is also written to the main memory. This is simple to design as memory is always up to date with respect to cache, but comes with the drawback that memory bandwidth is always consumed for writes.
Write Back Cache: In a write-back cache, every write operation to the cache is only written to cache. Write to main memory is deferred until the cache line is evicted or discarded from the cache. Write back Caches are better in terms of memory bandwidth as data is written back only when needed. The complexity comes in maintaining coherent data if there are multiple caches in system that can cache a same address, as memory may not always have latest data.

73. What is the difference between an inclusive and exclusive cache?

Inclusive and exclusive properties for caches are applicable for designs that have multiple levels of caches (example: L1, L2, L3 caches).
If all the addresses present in a L1 (Level 1) cache is designed to be also present in a L2 (Level 2) cache, then the L1 cache is called a strictly inclusive cache. If all the addresses are guaranteed to be in at-most only one of the L1 and L2 caches and never in both, then the caches are called exclusive caches.
One advantage of exclusive cache is that the multiple levels of caches can together store more data. One advantage of inclusive cache is that in a multiprocessor system, if a cache line has to be removed from a processor's cache, it has to be checked only in L2 cache while with exclusive caches, it has to be checked for presence in both L1 and L2 caches.

74. What are the different algorithms used for cache line replacement in a set-way associative cache?

Following are some of the algorithms that can be implemented for cache line replacements.
1) **LRU (Least Recently Used) Algorithm:** This algorithm keeps track of when a cache line is used by associating "age bits" along with cache line and discards the least recently used one when needed.
2) **MRU (Most Recently Used) Algorithm:** This is opposite to LRU and the line that is most recently used in terms of age gets replaced.
3) **PLRU (Pseudo LRU) Algorithm:** This is similar to LRU except that instead of having aging bits (which is costly with larger and higher associative caches), only one or two bits are implemented to keep track of usage.
4) **LFU (Least Frequently Used) Algorithm:** This algorithm keeps track of how often a line is accessed and decides to replace the ones that are used least number of times.
5) **Random replacement:** In this algorithm, there is no information stored and a random line is picked when there is a need for replacement.

75. What is the problem of cache coherency?

In Shared Multiprocessor (SMP) systems where multiple processors have their own caches, it is possible that multiple copies of same data (same address) can exist in different caches simultaneously. If each processor is allowed to update the cache freely, then it is possible to result in an inconsistent view of the memory. This is known as cache coherency problem. For example: If two processors are allowed to write value to a same address, then a read of same address on different processors might see different values.

76. What is the difference between snoop based and directory based cache coherency protocol?

Following is the difference between the two types of cache coherency protocols:
1) **Snoop based Coherence Protocol:** In a Snoop based Coherence protocol; a request for data from a processor is send to all other processors that are part of the shared system. Every other processor snoops this request and sees if they have a copy of the data and responds accordingly. Thus every processor tries to maintain a coherent view of the memory
2) **Directory based Coherence Protocol:** In a Directory based Coherence protocol; a directory is used to track which processors are accessing and caching which addresses. Any processor making a new request will check against this directory to know if any other agent has a copy and then can send a point to point request to that agent to get the latest copy of data.

Following are some of the advantages or disadvantages of each protocol:

Snoop Based Coherence	Directory Based Coherence
For smaller systems, Snoop based coherence will be faster if enough bandwidth for messages available	Directory based coherency can have longer latencies with a central lookup table that needs to be looked up before sending messages
Not scalable for larger SMP systems as messages need to be broadcasted for every request and can flood the system	Scales better and are used in larger SMP systems as there is no broadcast messages

77. What is a MESI protocol?

The MESI protocol is the most commonly used protocol for cache coherency in a design with multiple write back caches. The MESI stands for states that are tracked per cache line in all the caches and are used to respond to snoop requests. These different states can be explained as below:
1) **M (Modified)**: This state indicates that the cache line data is modified with respect to data in main memory and is dirty.
2) **E (Exclusive)**: This state indicates that the cache line data is clean with respect to memory but is exclusively present only in this memory. The exclusive property allows the processor in which this cache is present to do a write to this line
3) **S (Shared)**: This state indicates that the cache line data is shared in multiple caches with same value and is also clean with respect to memory. Since this is shared with all caches, the protocol does not allow a write to this cache line.
4) **I (Invalid)**: This state indicates that the cache line is invalid and does not have any valid data.

A cache can service a read request when the cache line is in any state other than Invalid. A cache can service a write request only when the cache line is in Modified or Exclusive state.

78. What are MESIF and MOESIF protocols?

These are two extensions of MESI protocol which introduces two new states "F" and "O" which are explained below:
1) **F (Forward)**: The F state is a specialized form of the S state, and indicates that a cache should act as a designated responder for any requests for the given line by forwarding data. If there are multiple caches in system having same line in S state, then one of them is designated as F state to forward data for new requests from a different processor. The protocol ensures that if any cache holds a line in the S state, at most one (other) cache only holds it in the F state. This state helps in reducing traffic to memory as without F state, even if a cache line is in S state in multiple caches, none of them cannot forward data to a different processor requesting a read or write. (Note that an S state line in cache can only service the same processors reads).
2) **O (Owned)**: The O state is a special state which was introduced to move the

modified or dirty data round different caches in the system without needing to write back to memory. A line can transition to O state from M state if the line is also shared with other caches which can keep the line in S state. The O state helps in deferring the modified data to be written back to memory until really needed.

79. What is a RFO?

RFO stands for Read for Ownership. It is an operation in cache coherency protocol that combines a read and invalidate broadcast. It is issued by a processor trying to write into a cache line that is in the Shared or Invalid states. This causes all other processors to set the state of that cache line to I. A read for ownership transaction is a read operation with intent to write to that memory address. Hence, this operation is exclusive. It brings data to the cache and invalidates all other processor caches that hold this memory address.

80. What is the concept of Virtual memory?

Virtual memory is a memory management technique that allows a processor to see a virtual contiguous space of addresses even if the actual physical memory is small. The operating system manages virtual address spaces and the assignment of memory from secondary device (like disk) to physical main memory. Address translation hardware in the CPU, often referred to as a memory management unit or MMU, translates virtual addresses to physical addresses. This address translation uses the concept of paging where a contiguous block of memory addresses (known as page) is used for mapping between virtual memory and actual physical memory. Following diagram illustrates this concept.

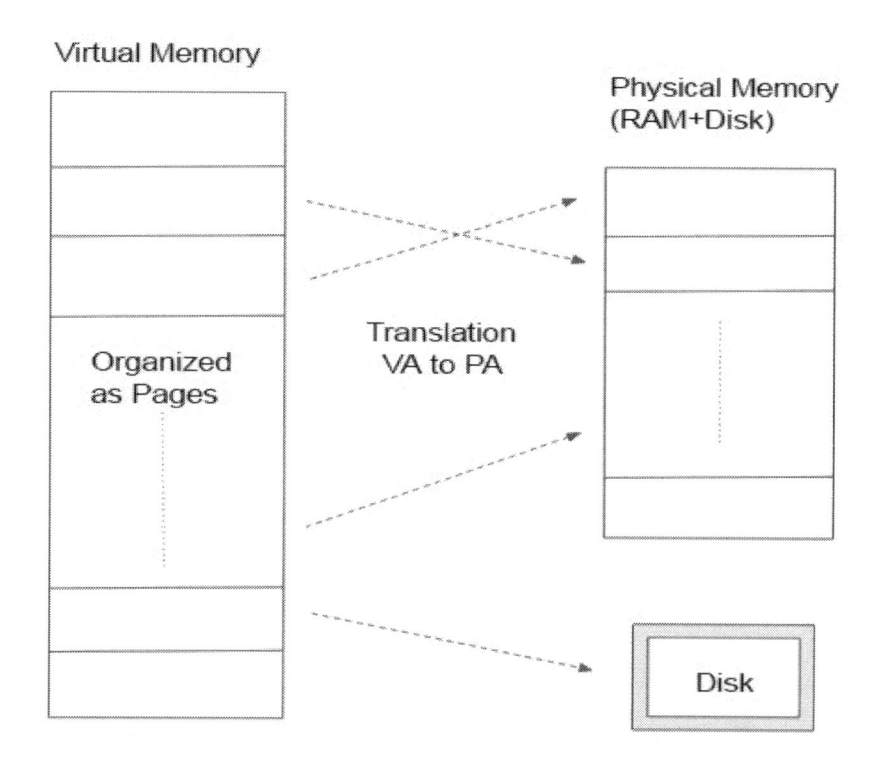

81. What is the difference between a virtual memory address and a physical memory address?

The address used by a software program or a process to access memory locations in it's address space is known as **virtual address**. The Operating System along with hardware then translates this to another address that can be used to actually access the main memory location on the DRAM and this address is known as **physical address**. The address translation is done using the concept of paging and if the main memory or DRAM does not have this location, then data is moved from a secondary memory (like Disk) to the main memory under OS assistance.

82. What is the concept of paging?

All virtual memory implementations divide a virtual address space into pages which are blocks of contiguous virtual memory addresses. A page is the minimum granularity on which memory is moved from a secondary storage to physical memory for managing virtual memory.
Pages on most computer systems are usually at least 4 kilobytes in size. Some architectures also supports large page sizes (like 1MB or 4MB) when there is a need of much larger real memory.
Page tables are used to translate the virtual addresses seen by the application into physical addresses. The page table is a data structure used to store the translation details of a virtual address to a physical address for multiple pages in the memory.

83. What is a TLB (Translation lookaside buffer)?

A TLB is a cache that stores the recent address translations of a virtual memory to physical memory which can be then used for faster retrieval later. If a program requests a virtual address and if it can find a match in the TLB, then the physical address can be retrieved from the TLB faster (like a cache) and the main memory need not be accessed. Only, if the translation is not present in TLB, then a memory access needs to be performed to actually do a walk through the page tables for getting the address translation which takes several cycles to complete. Following diagram illustrates this, where if the translation is found in the TLB, the physical address is available directly without needing to go through any Page table translation process.

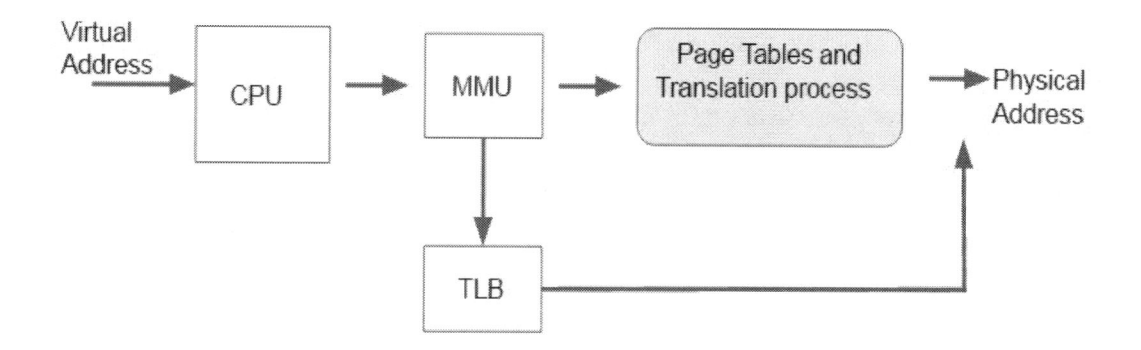

84. What is meant by page fault?

When a memory page that is mapped into Virtual Address Space but is not loaded into the main memory is accessed by a program, computer hardware [Memory Management Unit (MMU)] raises an interrupt. This interrupt is called Page Fault.

85. If a CPU is busy executing a task, how can we stop it and run another task?

The program execution on a CPU can be interrupted by using external interrupt sources.

86. What are interrupts and exceptions and how are they different?

Interrupt is an asynchronous event that is typically generated by an external hardware (an I/O device or other peripherals) and will not be in sync with instruction execution boundary. For example: An interrupt can happen from a keyboard or a storage device or a USB port. Interrupts are always serviced after the current instruction execution is over, and the CPU jumps to execution of the Interrupt service routine.

Exceptions are synchronous events generated when processor detect any predefined condition while executing instructions. For example: when a program encounters a divide by zero or an undefined instruction, it can generate an exception. Exceptions are further divided into three types and how the program flow is altered depends on the type:
1) **Faults:** Faults are detected and serviced by processor before the faulting instruction
2) **Traps:** Traps are serviced after the instruction causing the trap. The most common trap is a user defined interrupt used for debugging.
3) **Aborts:** Aborts are used only to signal severe system problems when execution cannot continue any longer.

87. What is a vectored interrupt?

A vectored interrupt is a type of interrupt in which the interrupting device directs the processor to the correct interrupt service routine using a code that is unique to the interrupt

and is sent by the interrupting device to the processor along with the interrupt.
For non-vectored interrupts, the first level of interrupt service routine needs to read interrupt status registers to decode which of the possible interrupt sources caused the interrupt and accordingly decide which specific interrupt service routine to be executed.

88. What are the different techniques used to improve performance of instruction fetching from memory?

1) Instruction Cache and Pre-fetch: An instruction cache and Prefetch algorithm will keep on fetching instructions ahead of the actual instruction decode and execute phases, which will hide the memory latency delay for instruction fetch stage in the design.
2) Branch Prediction and Branch Target Prediction: A Branch Prediction will help in predicting if a conditional branch will take place or not based upon the history, and A Branch Target Prediction will help predicting the target before the processor computes. This helps in minimizing instruction fetch stalls as the fetch algorithm can keep fetching instructions based on prediction.

89. What is meant by a superscalar pipelined processor?

A superscalar pipelined design uses instruction level parallelism to enhance performance of processors. Using this technique, a processor can execute more than one instruction during a clock cycle by simultaneously dispatching multiple instructions to different execution units on the processor. If the processor can execute "N" instructions parallely in a cycle then it is called N-way superscalar.

90. What is the difference between in-order and out-of-order execution?

In-Order Execution: In this model, instructions are always fetched, executed and completed in the order in which they exist in the program. In this mode of execution, if one of the instructions stalls, then all the instructions behind it also stall.
Out-of-Order Execution: In this model, instructions are fetched in the order in which they exist in the program, their execution can happen in any order, and their completion again happen in-order. The advantage of this model is that if one instruction stalls, then independent instructions behind the stalled instruction can still execute, thereby speeding up the overall execution of program.

91. What is the difference between a conditional branch and unconditional branch instruction?

A branch instruction is used to switch program flow from current instruction to a different instruction sequence. A branch instruction can be a conditional branch instruction or an unconditional branch instruction.
Unconditional Branch Instruction: A branch instruction is called unconditional if the

instruction always results in branching.
Example: `Jump <offset>` is an unconditional branch as the result of execution will always cause instruction sequence to start from the <offset> address
Conditional Branch Instruction: A branch instruction is called conditional if it may or may not cause branching, depending on some condition.
Example: `beq  ra , rb, <offset>` is a conditional branch instruction that checks if two source registers (ra and rb) are equal, and if they are equal it will jump to the <offset> address. If they are not equal, then the instruction sequence will continue in the same order following the branch instruction.

92. What is branch prediction and branch target prediction?

A branch predictor is a design that tries to predict the result of a branch so that correct instruction sequences can be pre-fetched into instruction caches to not stall instruction execution after encountering a branch instruction in the program. A branch predictor predicts if a conditional branch will be taken or not-taken.
A branch target predictor is different and predicts the target of a taken conditional branch or an unconditional branch instruction before the target of the branch instruction is computed by the execution unit of the processor.

93. What is meant by memory mapped I/O?

Memory Mapped I/O (MMIO) is a method of performing input/output (I/O) between a CPU and an I/O or peripheral device. In this case, the CPU uses the same address bus to access both memory and I/O devices (the registers inside I/O device or any memory inside the device). In the system address map, some memory region is reserved for the I/O device and when this address is accessed by the CPU, the corresponding I/O devices that monitor this address bus will respond to the access.
For Example: if a CPU has a 32 bit address bus: it can access address from 0 to 2^{32}, and in this region, we can reserve addresses (say from 0 to 2^{10}) for one or more I/O devices.

94. What's the advantage of using one-hot coding in design?

In one-hot coding, two bits are changing each time, one being cleared and the one being set. The advantage being that you don't need to do decode to know which state you are in. It uses more Flip-Flops but less combinational logic and in timing critical logic not having the decode logic may make the difference.

Chapter 3: Programming Basics

In the present day scenario where Digital VLSI Designs are trending towards SOC designs with increased complexity, the Design Verification job is visibly becoming more and more software oriented. Both Design and Verification Engineers need to possess strong hardware and software skills surrounding the hardware and software interface. Additionally, with verification testbenches and simulation models becoming more and more complex, software programming knowledge has become a must-have skill for any Verification Engineer.

A Design Verification Engineer should know at-least one programming language thoroughly (C/C++/SystemVerilog), should have strong scripting skills, working knowledge and familiarity with UNIX/Linux environments, and a good understanding of object oriented programming concepts.

Due to these reasons, questions that test a candidate's understanding of programming basics are part of almost all Verification job interviews. Ability to think and code algorithms, and model design behaviors forms a vital component of an interview.

This section is therefore organized into various sub-sections, and lists down some of the most commonly asked interview questions with their answers. First section consists of questions that test Basic Programming Concepts. Second section consists of questions on Object Oriented Programming, and the third section consists of programming examples specific to most commonly used languages. Third section is further sub-divided into three parts and has questions covering UNIX, Linux, C, C++, and Perl.

For better understanding, wherever possible, we have tried to explain the approach required to solve a problem.

3.1 Basic Programming Concepts

95. What is the difference between a static and automatic variable, local variable and a global variable in any programming language?

There are two separate concepts that distinguish local, auto, static and global variables. The "scope" of a variable defines where it can be accessed and the "storage duration" determines how long the variable can be accessed.

1) The scope of variable distinguishes between local and global variables. Local variables have limited scope and are visible only within the block of code in which they are declared. Global variables are accessible anywhere in the program once declared.

2) The storage duration distinguishes between an automatic and static variable. Static variables have a lifetime that lasts till the end of program and hence are accessible throughout. The scope is still defined based on whether it is a local or global variable.

Automatic variables are those which only have a lifetime until the program execution leaves the block/scope in which they are defined.

For Example: In following SystemVerilog code:
global_int is declared as a class member and has a global scope throughout the class while its life ends when the object of the class is de-referenced
global_static variable is declared as a static variable and has a global scope throughout the class as well as a lifetime throughout the program, even after the object of class is de-referenced or even constructed
sum variable is local to the function compute() and is visible only inside function and also exists only for the duration when compute() is executed
"count" variable is local to the function compute() and is visible only inside the scope of function but since it is static it only has a single copy and retains the value even after the function compute() is executed multiple times.

```
class test_class;
  int global_int; //automatic by default
  static global_static; //global static variable
  void function compute()
  begin
    static int count; //local static variable
    local int sum; //local automatic variable
    sum = sum +1;
    count = count +sum;
  end
endclass
```

96. What do you mean by inline function?

An inline function is a function that is expanded inline when invoked i.e. the compiler replaces the function call with the corresponding function code. This is beneficial if the function is a small code and used at several places. Instead of having the overhead of calling functions and returning from function, this will be faster (especially if the function is smaller).
For Example: In C, you can define an inline function called "max" as below and every instance of the function called inside main will be replaced with code instead of function call

```
inline int max(int a, int b) {
  return a > b ? a : b;
}

main () {
  int   a1,a2,a3,b1,b2,b3;
  int c1,c2,c3;
  c1 = max(a1,b1);
  c2 = max(a2,b2);
```

```
    c3 = max(a3,b3);
}
```

97. What is a "regular expression"?

A regular expression is a special sequence of characters that help a user match or find other strings (or sets of strings), using a special syntax. It is one of the most powerful concepts used for pattern matching within strings. Regular expressions are widely used in languages like Perl, Python, Tcl etc.

98. What is the difference between a heap and a stack?

A **stack** is a special region of memory that stores temporary variables created by a function. Every time a function declares a new automatic variable, it is pushed to the stack and every time the function exits, all of the variables pushed on the stack are deleted. All local variables uses stack memory. Stack memory is managed automatically and also has a size limitation. If stack runs out of memory, we get stack overflow error.
A **heap** on the other hand is a region of memory that needs to be managed. The programmer (and in some languages that support garbage collection, the compiler) has to allocate and free memory. These are typically used to store static variables as well as for objects. It is slightly slow compared to stack and is referenced through pointers. Also unlike stack, variables stored on heap can be referenced from anywhere in the program. Size of heap can be changed. Heap can have fragmentation problem when available memory is stored as disconnected blocks.

99. What is the difference between ++a and a++ in any programming language?

++a first increment the value of "a" and then returns a value referring to "a". Hence, if "++a" is assigned to a variable, then incremented value of "a" will be used.
a++ first returns value of "a" (which is current value of "a") and then increments "a". Hence, if "a++" is assigned to a variable, then old value of "a" will be used in the assignment.

100. What is meant by memory leak?

When we allocate memory dynamically but somehow lose the way to reach that memory, then it is called a memory leak. In certain programming languages like C++, every memory allocation done (say for creating objects), should be freed up (by calling destructors) without which those memories would leak and would no longer be available. In certain other languages like SystemVerilog, Java, etc., the language internals takes care of cleaning up memory and has lesser chance of memory leak.

101. What is the difference between a Compiler and an Interpreter?

For a machine (like a computer) to understand a code, it should be in a binary (combination of 0's and 1's) format. This binary code which is understandable by a machine is called "machine code".

Programmers usually write a computer program/code using a high-level programming language (e.g. C/C++/Perl/Python). High level computer program/code written is called source code. For a machine to understand this source code, it should be converted to a machine code. Compiler and Interpreters are programs that convert a source code into a machine code.

Compiler	Interpreter
Scans through entire program and translates **entire source code** to machine code.	Scans and translates source code taking **one statement** at time.
Requires large amount of time to analyze source code	Requires less amount of time to analyze source code
Outputs machine specific binary code.	Output code is some sort of intermediate code which is interpreted by another program
Faster Execution (as computer hardware executes it)	Slower Execution (as it is executed by another program)
Programming language like C, C++ use compilers.	Scripting language like Perl, Python use interpreters.
Generates an error message only after scanning the whole program.	Continues translating the program until the first error is met. Stops after first error is met.

102. What is the difference between a statically typed language and a dynamically typed language?

Statically Typed Language: A statically typed language is one in which types are fixed at compile time. This means you need to declare all variables with their datatypes before using them. For Example: Java, C, and SystemVerilog are statically typed languages.

Dynamically Typed Language: A dynamically typed language is one in which types are discovered at execution time. This is opposite to that of statically typed languages. For Example: VBScript and Python are dynamically typed and you need not declare all variables with their data types before using. They figure out what type a variable is when you first assign a value for the variable.

103. Which of the following is false related to Stack?

1) Only Push and POP operations are applicable to stack
2) Stack Implements a FIFO
3) Stack is useful for nested loops, subroutine calls
4) Stack is efficient for arithmetic expression evaluation

Option (2) is false.
Stack doesn't implement a FIFO but it is a LIFO (Last in First Out). Variables that are pushed onto stack last are popped back.

104. What are the major difference(s) between "use" and "require" in Perl?

Following are the major differences:
1) "use" is evaluated at compile-time, whereas "require" is evaluated at run-time.
2) "use" implicitly calls the import function of the module being loaded, whereas "require" doesn't.
3) "use" only expects a bareword, whereas "require" can either take a bareword or an expression.

105. What is the difference between static and dynamic memory allocation?

Static Memory Allocation	Dynamic Memory Allocation
Memory is allocated at compile time.	Memory is allocated at run time.
Memory is allocated either on a stack or on other sections of the program.	Memory is allocated on heap.
No need to free this memory. Lifetime of a variable in static memory is the lifetime of the program.	Need to explicitly free the memory.
Fixed in size. Once allocated, memory size can't be changed.	Can change in size. (using realloc())
Faster Execution	Slower Execution

106. What are the pre-processor directives?

In a code, Pre-processor directives are the lines that start with a hash (#) sign. These act as directives to a pre-processor that examines the code before the compilation of the code begins. Pre-processor changes the source code and the result is a new source code with these directives replaced. For Example: Normal syntax of a pre-processor directive is:
`#define identifier value`
Whenever pre-processor encounters "`identifier`" in the source code, it replaces it with "`value`", and generates a new source code before compilation.

107. What is the function of "using namespace std" in C++ codes?

A "namespace" allows a programmer to group entities like classes, objects, and functions under a name. "std" is short for the word "standard". A standard namespace (std

namespace) is a special type of namespace where all built-in C++ library routines are kept (like string, cin, cout, vector etc.).
Therefore, "using namespace std" tells C++ compiler to use the standard C++ library.

108. What is the difference between following initializations: "int a;" and "const int a;"

`const` keyword tells the compiler that a variable or object should not change once it has been defined. Once defined, a const variable/object is not assigned to any other value in any way during program duration.
Hence, if "a" is declared as "int a", value of a can be changed at different times in the program.
However, if "a" is declared as "const int a", once initialized it cannot be changed again.

109. What does the keyword "volatile" mean in C language?

"volatile" keyword in C tells the compiler that the value of a variable (declared as volatile) may change between different accesses and hence it directs the compiler not to optimize anything related to the volatile variable. volatile keyword is mainly used while interfacing with memory mapped Input Output (Hardware). Once a variable is declared as volatile, compiler can't perform any optimizations like: removing memory assignments, caching variables in registers, or changing the order execution of assignments.

110. What is a pointer? Explain the concept of pointer.

Pointer is a variable whose value is an address of another variable. A pointer contains direct address of a memory location. The asterisk sign * is used to denote a pointer.
`int *p;` tells the compiler that variable "p" is a pointer, whose value is an address of a memory location where an integer variable is stored.
Similarly, `float *f;` tells the compile that variable "f" is a pointer, whose value is an address of a memory location where a float variable is stored.

To understand this in detail, consider following snippet of a C code:
```
int a = 10;
int *b;
int c;
b = &a;
c = *b;
printf("b=%d and c=%d\n",b,c);
```

Here, value of variable "a" is 10. "&a" denotes memory address where value of the variable "a" is stored. Let us assume that variable "a" is stored at a memory location "0x1000" (4096 in decimal).

Since pointer "b" is assigned address of "a", value of variable "b" would be "4096". Further, variable "c" would be equal to value of variable "a". Therefore, output of above printf would be: `b=4096` and `c=10`.
Diagram below represents this example:

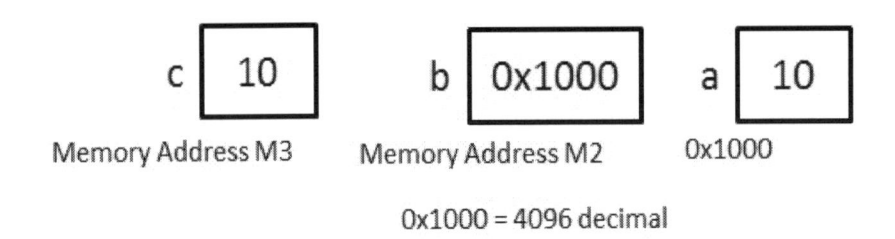

```
c | 10              b | 0x1000           a | 10

Memory Address M3   Memory Address M2      0x1000

            0x1000 = 4096 decimal
```

111. Explain the concept of "Pass by Value" and "Pass by Reference" in C?

Pass by Value: In this case, a copy of the "Value" of actual parameter is made in the memory, and is assigned to a new variable. This means that any changes made to the value of the new variable inside the function are local to that function and doesn't impact original value of the actual parameter.

Pass by Reference: In this case, a copy of the "Address" of actual parameter is made in the memory and is assigned to a new variable. This is used when the function makes changes to the variable which will then also reflect in the original parameter value. Parameters that normally consume more memory are normally passed by reference to avoid creating copies local to function. Example: arrays are normally passed by reference.

For Example: In the code below, we have a function named ""pass_by_value", where we pass actual value stored in variable "c" (15) to a local variable (named "a") in the function. In this case, a new memory location will be assigned for variable "a" and value of "15" will be stored in that memory location. Any changes made to value of "a" inside the function would ONLY change the value of variable "a", and since "a" is stored at a different memory location than "c", value of variable "c" won't change. Hence, output of first printf in the program will be "`1. Value of C is 15`". This is concept of Pass by Value.

Also, in the code below, we have a function named "pass_by_reference", to which we pass the memory address of the location where value of variable "c" is stored. Here, we use an integer pointer (int *b) to capture the memory location where value of an integer variable "c" is stored, and inside the function we tell the compiler to add "5" to the value present at the memory location captured in b (*b), which is nothing but value of variable "c" (as variable "c" is stored at that memory location). Hence, output of second printf in the program will be "`2. Value of C is 20`". This is concept of Pass by Reference.

```c
#include<stdio.h>
void pass_by_value(int a){
   a = a+5;
}
```

```
void pass_by_reference(int *b){
  *b = (*b)+5;
}

int main(){
  int c=15;
  pass_by_value(c);
  printf("1. Value of C is %d\n",c);

  pass_by_reference(&c);
  printf("2. Value of C is %d\n",c);
  return 0;
}
```

112. What is the Value and Size of a NULL pointer?

NULL pointer can be defined as: `int *a = NULL;`
Value of a NULL pointer is 0.
As we have seen through previous questions/answers, pointer is a variable whose value is an address of another variable. Since, value of a pointer is an address; size of a pointer would vary depending upon the machine. If it is 32-bit machine, pointer size would be 4 bytes and if the machine size is 64-bit, size of pointer would be 8 bytes.

113. What is a linked list? How many different types of linked list are there?

A Linked List is a data structure consisting of a group of nodes which together represent a sequence. In a simplest form, each node is composed of data and a reference (link) to the next node in the sequence. Following are different types of linked lists:
   1) Linear Linked List or One Way Linked List or Singly Linked List.
   2) Two Way Linked List or Doubly Linked List.
   3) Circular Linked List.

114. Of what order is the "Worst case" Time complexity of following algorithms?
   1) Linear Search
   2) Binary Search
   3) Insertion Sort
   4) Merge Sort
   5) Bucket Sort

Time complexity in simple terms can be thought of as: "How long does a particular algorithm run?". It is amount of time taken by an algorithm to run as a function of input string length.

It makes a lot of sense to be able to estimate run time before starting an algorithm/program to see if we are efficiently using all the resources. If a program/algorithm takes lot of time to execute, we would be consuming a lot more machine resources, which is costly.

Time complexity of an algorithm is expressed using *Big O* notation. In *Big O* notation we exclude coefficients and lower order terms. For Example: If time complexity of an algorithm is calculated to be equal to $5n^4 + 6n^2 + 1$, using Big O notation, we say that time complexity of the algorithm is $O(n^4)$ [i.e. Order of n^4]

Since performance of an algorithm may vary with different inputs of same size, time complexity can be of three types: Best Case, Average Case and Worst Case. We are often not interested in the best case time complexity. We usually tend to compute "average case" or "worst case" time complexity.

For the common algorithms asked in this question, following are the worst case time complexities:

1) O(N)
2) O(log(N))
3) O(N²)
4) O(N*log(N))
5) O(N)

Where, N = size of the list.

115. Of what order is the Space complexity of following algorithms?
1) Linear Search
2) Binary Search
3) Insertion Sort
4) Merge Sort
5) Bucket Sort

Concept of space complexity is similar to the concept to time complexity (mentioned above in previous answer) with the difference that: Space complexity represents the total amount of storage space (memory) required to execute a program or solve a problem with a particular algorithm. It is also represented in Big O notation.

1) O(1)
2) O(1)
3) O(N)
4) O(N)
5) O(N)

Where, N = size of the list.

116. Explain the difference between "&" and "&&" operators in C/C++?

& is a bitwise AND operator while && is a logical AND operator.
Logical operators work with boolean values - true (1) and false (0), and return a boolean value.
Bitwise operators perform bit operation on each of the bit and return a bit value.

Bitwise operator usage: if a =10 and b = 6, a & b will return 2 (4'b1010 & 4'b0110 = 4'b0010)
Logical Operator: If a=10 and b=6, then the following expression will return true as it operates on two boolean values which are true in this example.
```
c = (a==10) && (b==6);
```

117. How does a "Structure" and a "Union" differ in terms of memory allocation in C/C++?

Structs allocate enough space to store all of the fields/members in the struct. The first one is stored at the beginning of the struct, the second is stored after that, and so on. Unions only allocate enough space to store the largest field listed, and all fields are stored at the same space. This is because in a union, at a time only one type of enclosed variable can be used unlike struct where all the enclosed variables can be referenced.

118. How much memory is allocated for struct ID below?
```
struct ID {
  int IntID;
  char CharID[8];
};
```
12 Bytes (4 bytes are allocated for an integer and 8 Bytes for character array)

119. How much memory is allocated for union ID below?
```
union ID {
  int IntID;
  char CharID[8];
};
```
8 Bytes (8 Bytes for character array) [For Unions, Memory is allocated only for the largest field]

120. What is a "Stack" and how does a "Stack" differ from "FIFO"?

Stack is a data structure. Stack is "Last In First Out" whereas FIFO is "First In First Out".

121. What is a kernel?

A Kernel is a computer program that manages Input/output requests from software and translates these requests into CPU instructions or other electronic instructions for the computer.

122. What is UNIX and how is it different from Linux?

UNIX is basically an operating system (like windows). It is a multi-user, multi-tasking system for servers, desktops and laptops. It is made up of three parts: Kernel, Shell, and the Program. We have already seen what a kernel is (previous question). Shell is an interface between the user and the kernel.
Both Linux and UNIX are operating systems. Linux is a free and open source operating system, whereas different versions of UNIX have different costs based upon the vendors. Linux is developed by open source development. Ubuntu is an example of Linux distributor whereas Solaris is an example of UNIX distributor.

123. What does Perl stand for?

Practical Extraction and Reporting Language.

124. Perl is an interpreted language? What does this mean?

Perl is an interpreted language means that: it scans and translates the source code taking one statement at a time, where it converts source code into an intermediate code that can be interpreted by another program. It also means that it doesn't convert entire source code into a binary code (like a compiler does). Perl continue source code translation till it meets first error and stops further processing.

125. What are different types of data types in Perl?

Scalars: Preceded by $, scalars are simple variables. A scalar can be a number, or a string or a reference.
Arrays: Preceded by @, arrays are ordered lists of scalars. Index for arrays start from 0.
Hashes: Preceded by %, hashes are unordered sets of keys/value pairs that can be accessed using keys as subscripts.

126. What is a Cron Job? When to use a Cron Job?

A Cron Job is a time based job scheduling in an operating system. It allows you to automatically run your jobs periodically at specified times, dates, days, intervals, etc.
For Example: Say a user has a Shell or a Perl script that calculates per-person disk space usage for a disk in UNIX/Linux. Setting a Cron job in UNIX/Linux for this script with specified frequency (or time) would ensure that this script runs automatically at scheduled times (or frequencies) without the user having the need to manually run it every time.

127. What is the use of "rsync" command in UNIX/Linux?

"rsync" stands for "Remote Sync", and it is a commonly used command for copying or synchronizing files/directories across disks, networks, servers, and machines. rsync minimizes the amount of data required to be copied as it moves only those portions of the files that have changed. "rsync" also consumes less bandwidth as it uses some compression and decompression methods while sending and receiving the data. One of the most common use of "rsync" command is to perform data backup and mirror disks between two machines.

128. Variable names in C can contain alphanumeric characters as well as special characters? True or False?

False

129. What is the use of a '\0' character in C/C++?

It is terminating null character and is used to show the end of a string.

130. What are binary trees?

Binary trees are an extension of the concept of linked lists. A binary tree has two pointers: "a left one" and "a right one". Each side can further branch to form additional nodes which each node having two pointers as well.

131. How can you generate random numbers in C?

Using `rand()` function and including `<stdlib.h>` header file.
In general terms, concept of random number generation is important for testing a particular program/code for wide range of inputs to see if the code is working properly for various different inputs. `rand()` help randomizing input conditions and generating random inputs for testing a code.

132. What are Special Characters, Quantifiers, and Anchors w.r.t regular expressions?

Special Characters are meta-characters that give special meaning to the syntax of regular expression search. Example: \, ^, $, (), [], |, &
Quantifiers specify "how often" we match a preceding regular expression. Example: *, +, ?, {}
Anchors specify "where" we match. Anchors allow a user to specify position for text/pattern search. Example: ^, $, <, >

3.2 Object Oriented Programming Concepts

133. What is the difference between a class and object?

A class is a set of attributes and associated behaviour that can be grouped together. An object is an instance of the class which then represents a real entity with attributes and behaviour. The attributes can be represented using class data members while the behavior can be represented using methods.
For Example: An animal can be represented as a class while different animals like dog, cat, etc., can be objects of animal class.

134. What is the difference between a Class and a Struct in C++?

A "struct" was originally defined in C to group different data-types together to perform some defined functionality. However, in C++ this grouping was extended to include functions as well. A "class" is also a data type that can group different data types and functions operating on a defined functionality.
The only real difference in C++ is that all members of a class are private by default, whereas all members of a struct are public by default.

135. What is the difference between a Class and a Struct in SystemVerilog?

In SystemVerilog, both class and struct are used to define a bundle of data types based on some functionality to be performed. However, a struct is an integral type and when it is declared, the necessary memory is allocated. On the other hand, a class is a dynamic type and once you declare a class, you only have a class handle referenced to null. Memory allocation happens only when an actual object of the class is created.

136. What are public, private and protected members?

These are different access attributes for class members.
1) Private data members of a class can only be accessed from within the class. These data members will not be visible in derived classes
2) Public members can be accessed from within the class as well as outside the class also. These are also visible in derived classes
3) Protected data members are similar to private members in the sense that they are only accessible within the class. However unlike private members, these are also visible in derived class.

137. What is Polymorphism?

Polymorphism means the ability to assume several forms. In OOP context, this refers to the ability of an entity to refer to objects of various classes at run time. This is possible with the concept of inheritance and virtual functions in SystemVerilog (and with the concept of function and operator overloading, which are present in C++). Depending on the type of object, appropriate method will be called from the corresponding class.

138. What are Method Overriding and Method Overloading? What is the difference between the two?

Method overloading is the ability of the functions with same names to be defined multiple times with different set of parameters.
For Example: A function add() can be defined in three formats as shown below. Based on the type and number of arguments passed when the function is called, the correct definition will be picked up.

```
1) function add (int operand1, int operand2);
2) function add (int operand1, int operand2, int operand3)
3) function add (float operand1, float operand2)
```

Method overriding is the ability of the inherited class redefining the virtual method of the base class. Method overriding is supported in most of the object oriented programming languages.

However, unlike C++, method overloading is not supported in SystemVerilog language. SystemVerilog language only supports method overriding in terms of virtual methods and derived classes. For Example: A base class can define a function called compare() which compares two data members of its class as follows

```
class BaseClass;
  int a, b;
  virtual bit function compare();
    if(a==b) return 1;
  endfunction
endclass
```

A derived class can override this definition based on new data members as shown below:

```
class DerivedClass extends BaseClass;
  int c;
  function compare();
    if((a==b) && (a==c)) return 1;
  endfunction
endclass
```

139. What is operator overloading?

In object oriented programming, operator overloading is a specific case of polymorphism, where different built-in operators available can be redefined or overloaded. Thus a programmer can use operators with user-defined types as well.

This is supported in C++ while not supported in SystemVerilog.

Following examples shows a Testclass where the operator + is overloaded such that two class objects of type "Testclass" can be added. The implementation then adds the data members from two objects and assigns it to the data member of result class.

```cpp
#include <iostream>
class Testclass{
public:
   int a;
   int b;
   Testclass operator+(const Testclass& obj);
}

Testclass Testclass::operator+(const Testclass& obj2){
   Testclass tmp_obj = *this;
   tmp_obj.a = tmp_obj.a + obj2.a;
   tmp_obj.b = tmp_obj.b + obj2.b;
   return tmp_obj;
}

int main(void){
   Testclass obj1, obj2, obj3;
   obj1.a = 1;
   obj1.b = 1;
   obj2.a = 2;
   obj2.b = 2;
   obj3.a = 0;
   obj3.b = 0;
   obj3 = obj1 + obj2;
   std::cout<<obj3.a<<"   "<<obj3.b<<"\n";
   return 0;
}
```

140.   What is a constructor method?

Constructor is a special member function of a class, which is invoked automatically whenever an instance of the class is created. In C++, it has the same name as its class. In SystemVerilog, it is implemented as new() function.

141.   What is destructor?

Destructor is a special member function of a class, which is invoked automatically whenever an object goes out of the scope. In C++, it has the same name as its class with a tilde character prefixed while in SystemVerilog, there is no destructor as the language supports automatic garbage collection.

142. What is the difference between composition and inheritance in OOP?

Composition uses a "has - a" relationship between two classes. When a class instantiates object of another class, the relationship is "has-a" and this property is called composition. Inheritance uses a "is - a" relationship between two classes. When a class derives from another class, the relationship is a "is-a" and this property is called inheritance.
Following diagram illustrates this. If there is a base `class car` and a `class ford` is derived from this, then the relationship is "is - a", meaning the ford class is a car class. If the ford class has an object of engine class inside, then the relationship is HAS-A as shown in the diagram.

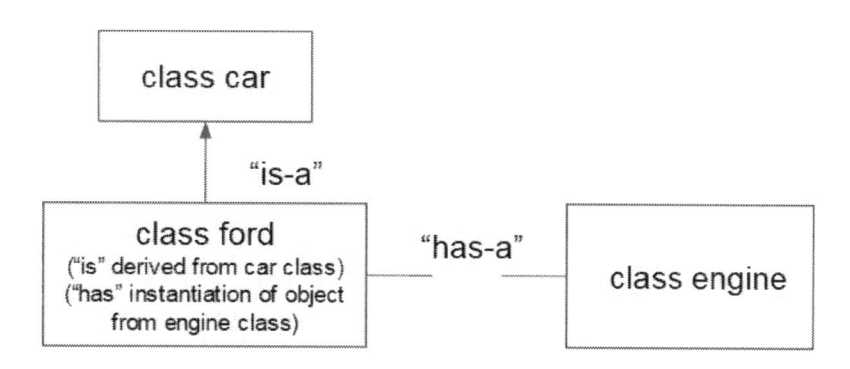

```
                    class car
                        ▲
                      "is-a"
   class ford
 ("is" derived from car class)     "has-a"
 ("has" instantiation of object  ----------   class engine
      from engine class)
```

143. What is the difference between a shallow copy and a deep copy used in object oriented programming?

In a Shallow copy, a new object is created that has an exact copy of the values as in the original object. If any of the fields of the object are references to other objects, just the reference addresses are copied (handles).
In a Deep copy, a new object is created that has exact copies of the values as in the original object. If any object has references to other objects, a copy of all values that are part of it are also copied and not just the memory address or handle. (Hence, known as deep copy)

For Example: Consider the following two classes A and B:
```
class A;
  int a;
  int b;
endclass

class B;
  int c;
```

```
    A    objA;
endclass
```

If a shallow copy method is implemented in class B, then when we copy B to a new object, only the memory handle of "objA" is copied over. In case of a deep copy, all values of A (namely its data members - a and b) are also copied and not a memory handle of "objA".

144. What are virtual functions in C++ or other OOP languages?

A virtual function is a member function that is declared within a base class and can be re-defined by a derived class. To create a virtual function, the function declaration in the base class is preceded by the keyword virtual. This way of re-defining a base class function in a derived class is also known as method overriding.

145. What is meant by multiple inheritance?

Multiple inheritance is a feature of some object-oriented computer programming languages in which an object or class can inherit characteristics and features from more than one parent object or parent class. It is distinct from single inheritance, where an object or class may only inherit from one particular object or class. Note: C++ supports multiple inheritance while SystemVerilog language doesn't.

146. What is an abstract class?

Abstract classes are classes that contain one or more abstract methods. An abstract method is a method that is declared, but contains no implementation. Abstract classes may not be instantiated, and require subclasses to provide implementations for the abstract methods. In SystemVerilog, the class name is prepended with a virtual keyword to make it an abstract class.

Following is an example of how an abstract class is defined with function defined as virtual. The derived classes can then actually implement this function.

```
virtual class  BaseShape;
  virtual function int get_num_edges();
  endfunction
endclass

class Rectangle extends BaseShape;
  function int get_num_edges();
    return 4;
  endfunction
endclass
```

147. What are static methods inside a class?

Static methods are methods defined inside a class using static keyword. These can be used without creating an object of the class. Also if there are multiple objects created of this class, there will still be only one static method which will be part of all objects.

148. What is 'this' pointer with reference to class?

this pointer is a special pointer that can be used to reference the current object of a class inside the class scope.

149. What is type conversion and type-casting in programming languages?

Type conversion and Type casting are different ways of changing one data type to another in a programming language.

A **type conversion** is an implicit conversion from one form to another, while a **type casting** is an explicit conversion.

For Example: if we have an int and a floating point variable as below and if we assign the floating point to an integer, the compiler implicitly does a type conversion.

```
int   a;
double b;
a = b;
```

In **type casting**, the programmer does an explicit conversion as shown below:

```
double a = 2.2
double b = 3.3;
int c = (int) a + (int) b;   //In this case the decimal values will
be truncated and we get a result of c=5
```

3.3 Programming questions

3.3.1 UNIX/Linux

150. How can you find out details regarding what a UNIX/Linux command does?

man <command-name>
Example: man grep

151. Write a UNIX/Linux terminal command: (assume filename = file.txt)
1) To display first 10 lines of a file
2) To display the 10th line of a file
3) To delete 13th line from a file
4) To delete last line from a file
5) To reverse a string (ex: "Hello")
6) To check if the last command was successful
7) To find number of lines in a file
8) To find number of characters in a file
9) To find number of characters on 17th line in a file
10) To get 3rd word of 17th line in a file
11) To change permission of a file to "Read" and "Executable" for all users.
12) To change group access permissions of a file to a group. (assume new group name as "new_group")
13) To move content to two files (file1.txt and file2.txt) into one file(file.txt)
14) To display all the processes running on your name
15) To uniquely sort contents of a file (file1.txt) and copy them to another file (file2.txt)
16) To check the username
17) To login to a remote host (say "remote-server")

1) Any of the following would work:
   a) head -10 file.txt
   b) cat file.txt | head -10
   c) sed "11,$ d" file.txt
2) head -10 file.txt | tail -1
3) sed -i "13 d" file.txt
4) sed -i "$ d" file.txt
5) echo "Hello" | rev
6) echo $?
7) cat file.txt | wc -l
8) cat file.txt | wc -c
9) head -17 file.txt | tail -1 | wc -c
10) head -17 file.txt | tail -1 | cut -f3 -d' '
11) chmod 555 file.txt
12) chgrp new_group file.txt
13) cat file1.txt file2.txt > file.txt
14) ps -aef
15) sort -u file1.txt > file2.txt
16) whoami
17) ssh username@remote-server

152. Write a UNIX/Linux terminal command to display following from a file (assume filename = file.txt):
1) All lines that contain the pattern "cat"
2) All lines that contain the word "cat"
3) All lines that doesn't contain the word "cat"

4) All lines that contain the word "cat" (case in-sensitive)
5) All lines that start with pattern "cat"
6) All lines that end with pattern "cat"
7) All lines that contain patterns "cat" and "123" (with pattern "cat" appearing before pattern "123")

1) grep "cat" file.txt
2) grep -w "cat" file.txt
3) grep -v -w "cat" file.txt
4) grep -i "cat" file.txt
5) grep "^cat" file.txt
6) grep "cat$" file.txt
7) grep "cat.*123" file.txt

153. Write a UNIX/Linux terminal command to list out names of all files in a directory (say /usr/bin/dir/) (and its subdirectories) that contain case insensitive pattern "I am preparing for Interview".

grep -ilr "I am preparing for Interview" /usr/bin/dir/*

154. A file (say /usr/home/file.txt) contains a list of directories. Write set of UNIX/Linux commands that looks at contents of this file, goes to each directory and runs a process (say script.pl). Assume that each line of the file (/usr/home/file.txt) contains path to only one directory.

foreach x (`cat /usr/home/file.txt`)
foreach> cd $x
foreach> script.pl
foreach> end

155. Write a UNIX/Linux terminal command that moves all non-blank lines from a file (file1.txt) to another file (file2.txt)

grep -v "^$" file1.txt > file2.txt

156. Write a UNIX/Linux terminal command to (assume filename = file.txt wherever applicable):
1) Find if a file exists in current directory or its sub-directories
2) Find if a file exists in a directory "/usr/bin/DIR" or its sub-directories
3) Find if a file exists in current directory only
4) Find if a file containing a specific word "dummy" in its name exists in current directory or its sub-directories
5) Find if a file with case insensitive name "file" exists in current directory or its sub-directories

6) Find all the files whose names are not "file.txt" and are present in current directory or its sub-directories
7) Rerun previously executed find command

1) find . -name "file.txt" OR find -name "file.txt"
2) find /usr/bin/DIR -name "file.txt"
3) find -maxdepth 1 -name "file.txt"
4) find . -name "*dummy*"
5) find . -iname "file"
6) find -not -name "file.txt"
7) ! find

157. Write a UNIX/Linux command to:
1) List all the Cron Jobs set on your name on a machine
2) List all the Cron Jobs set up by a user on a machine
3) Remove all the Cron Jobs set on your name on a machine
4) Remove all the Cron Jobs up by a user on a machine (If you have permissions to do so)
5) Edit a Cron Job on your name on a machine.
6) Set up a Cron Job that runs every day at 6:30PM
7) Set up a Cron Job that runs every minute.
8) Set up a Cron Job that runs first 20 days of every month at 6:30AM
9) Set up a Cron Job that runs only on a Friday every month at times 6:30AM and 6:30PM

1) crontab -l
2) crontab -u <user_name> -l
3) crontab -r
4) crontab -u <user_name> -r
5) crontab -e
6) 30 18 * * * <command_to_invoke_your_process>
7) * * * * * <command_to_invoke_your_process>
8) 30 6 1-20 * * <command_to_invoke_your_process>
9) 30 6,18 * * 6 <command_to_invoke_your_process> (assuming Sunday is represented by 0)

158. Mention shell hot keys that does following:
1) Kills a process
2) Moves a process running on a terminal to the background
3) Moves cursor to the beginning of a command at the shell
4) Moves cursor to the end of a command at the shell

1) Ctrl + c
2) Ctrl + z
3) Ctrl + a
4) Ctrl + e

3.3.2 Programming in C/C++

159. Write a C code to detect if underlying architecture in a machine is little Endian or big Endian.

Let's assume we have 32-bit machine.
If we have an unsigned integer 1, it would be stored in following format in Little Endian machine:

```
+----+----+----+----+
|0x01|0x00|0x00|0x00|
+----+----+----+----+
```
   (LSB Memory Address)   (MSB Memory Address)

And in following format in Big Endian machine:

```
+----+----+----+----+
|0x00|0x00|0x00|0x01|
+----+----+----+----+
```
   (LSB Memory Address)   (MSB Memory Address)

```c
#include <stdio.h>
int main()
{
  unsigned int i = 1;
  char *c = (char*)&i;
  if (*c)
    printf("Little Endian \n");
  else
    printf("Big Endian \n");
  return 0;
}
```

160. What will be the value of b and c in following code?
```
a = 10;
b = a++;
c = ++a;
```

b will have a value of 10 and c will have a value of 12.
The post-increment operator will make the increment only after assignment and hence b sees the value of a before increment. A pre-increment operator will first make the increment and hence a will be incremented from 11 (a changes to11 after assignment of b=a++) to 12

161. What would be the output of following C program?
```
#include<stdio.h>
int xyz=10;
int main() {
  int xyz=20;
  printf("%d",xyz);
  return 0;
}
```

The variable xyz is defined with both global and local scope. When printed in the function, the one with the local scope is printed and hence it will print a value of 20.

162. Find the Value of "y" in the following C code:
```
int main() {
  int x=4;
  float y = * (float *) &x;
  return 0;
}
```

Very Small Value. Some compilers may show answer as 0.
Following is the concept: "(float *) &x", tells compiler that pointer is to a float number stored at a memory location. Once we dereference this i.e. "* (float *) &x", this would imply: "value of a float number stored at a memory location". Floats are stored differently from integer (as for float numbers, bit [31] represents the signed bit, bits [30:23] represent the exponent and bits [22:0] represent the fraction). Hence 4.0 would become very-very small value when interpreted as a float (00000000000000000000000000000100).

163. What will be the output of the following C program?
```
#include<stdio.h>
int main()
{
  int i=0;
  for(i=0;i<20;i++)
  {
    switch(i) {
      case 0:i+=5;
      case 1:i+=2;
      case 5:i+=5;
      default: i+=4;
      break;
    }
    printf("%d\n",i);
  }
  return 0;
```

}

Answer:
16
21

This question is bit tricky. To understand the answer better, let us iterate over the "for" loop step by step. When the integer variable "i" is equal to 0, "case 0" will execute and it will change the value of integer variable "i" to 5. **Since, there is NO break statement** after "case 0", "case 1" will be executed and this will change the value of "i" to 7. Similarly "case 5" and "default" cases will also be executed and "i" will become 16 (5+2+5+4) when the first break statement after "default" case is encountered. Hence, "16" will be printed in the first iteration. Now, integer variable "i" will increment to "17" (as a result of i++ present as part of "for" loop). Since, "i" is 17, "switch" will choose "default" case and "i" will increment by "4" to become "21", and this will also be printed.

164. **Write a recursive function to find out factorial of a number "n" where n is always >=0.**

```
int factorial (int x)
{
  if ( (x==0) || (x==1) )
    return 1;
  else
    return (x*factorial(x-1));
}
```

165. **Generate a Fibonacci series using a recursive function.**

```
int fibonacci (int num)
{
  if( (num==0) || (num==1) )
    return num;
  else
    return (fibonacci(num-1) + fibonacci(num-2));
}
```

166. **What is the output of this C code when run on a 64-bit machine?**
```
#include <stdio.h>
int main()
{
  int x = 10000;
  double y = 56;
  int *p = &x;
  double *q = &y;
```

```
        printf("p and q are %d and %d", sizeof(p), sizeof(q));
        return 0;
}
```

p and q are 8 and 8

Since "p" and "q" are pointers, they are nothing but addresses in a 64-bit machine. Size of both would be 64 bits (8 bytes), irrespective of whether they point to an integer or a double data type.

167. What is a Linked List and when would you prefer to use linked lists?

As we have already seen in previous section, a Linked List is a data structure consisting of a group of nodes which together represent a sequence. In a simplest form, each node is composed of data and a reference (link) to the next node in the sequence.

Linked lists are preferred when we don't know the volume of data to be stored. For example: we can use linked lists in an employee management system, where we can easily add record of a new employee (addition of a new node - dynamic memory allocation), delete record of an old employee (removal of a node), edit record of an employee (editing data in a node).

For Questions 168 to 172: Using the variables and declarations given below:

```
struct node;
typedef struct node NODE;
typedef int Element;

// A pointer to a node structure
typedef NODE *LINK;

// A node defined as having an element of data
// and a pointer to another node
struct node {
  Element elem;
  LINK next;
};

// The Head or start of the List
typedef struct
{
  int size;
  LINK start;
} ListHead;
```

168. Write a C function to create a singly linked list.

For creating a singly linked list, we need to:
1. Create HEAD (h) of the linked list,
2. Initialize the size of linked list (to zero), and
3. Point the start pointer to NULL (as linked-list is empty at the time of creation).

Refer to following function for creating a singly linked list:

```
ListHead createList() {
  ListHead h;
  h.size = 0;
  h.start = NULL;
  return h;
}
```

169. Write a C function to Insert an Element at the head of a singly linked list.

When an element (e) has to be inserted at the HEAD of a linked list (h), we need to:
1. Dynamically allocate memory for a new NODE,
2. Assign value to the element in the new NODE,
3. Point the "next" pointer in the new NODE to the NODE which HEAD was previously pointing to, and
4. In the linked list HEAD, increment the "size" variable (as new NODE is added) and point the "start" pointer to the new NODE.

```
ListHead InsertElementAtHead(Element e, ListHead h) {
  LINK nl= (LINK) malloc (sizeof(NODE));
  nl->elem = e;
  nl->next = h.start;
  h.start= nl;
  h.size++;
  return h;
}
```

170. Write a C function to Insert an Element at the tail of a singly linked list.

When an element (e) has to be inserted at the tail of a linked list (h), we need to:
1. Dynamically allocate memory for a new NODE,
2. Assign value to the element in the new NODE,
3. Point the "next" pointer in the new NODE to NULL (as new NODE represents the tail of the linked list),
4. If Linked list is initially empty, point the "start" pointer in HEAD to new NODE, else traverse Linked List to find out the last NODE in the linked list and point the "next" pointer in the last NODE to new NODE.
5. Increment the "size" variable (as new NODE is added) in the linked list HEAD.

```
ListHead InsertElementAtTail(Element e, ListHead h) {
  LINK temp;
  LINK nl;
  nl=(LINK) malloc (sizeof(NODE));
  nl->elem=e;
  nl->next=NULL;
  if(h.start==NULL)
    h.start=nl;
  else
  {
    temp=h.start;
    while(temp->next!=NULL)
    temp=temp->next;
    temp->next=nl;
  }
    h.size++;
    return h;
}
```

171. Write a C function to Insert an Element at position "pos" in a singly linked list.

When an element (e) has to be inserted at a position (pos) in a linked list (h), we need to:
1. Dynamically allocate memory for a new NODE,
2. Assign value to the element in the new NODE,
3. If "pos" is more than the size of linked list, return an error message (as this is not possible). Else if "pos" is "0", insert the element at head (as seen above). Else, traverse through the linked list to the NODE just before "pos". Point "next" pointer in the new NODE to the NODE which NODE at "pos-1" was pointing to and point "next" pointer in NODE at "pos-1" to new NODE. Remember that "count" starts from 0.
4. Increment the "size" variable (as new NODE is added) in the linked list HEAD.

```
ListHead InsertAtPos(Element e, ListHead h, int pos) {
  LINK temp;
  LINK nl;
  nl=(LINK)malloc(sizeof(NODE));
  nl->elem=e;
  int count=0;
  if(pos>h.size) {
    printf("Error: Wrong position \n");
    return h;
  }
  if(pos==0) {
    nl->next=h.start;
    h.start=nl;
  }
```

```
    else {
      for (temp = h.start; count<(pos-2); temp = temp->next, count++)
;
      nl->next=temp->next;
      temp->next=nl;
    }
    h.size++;
    return h;
  }
```

172. Write a C function to Delete an Element from a singly linked list.

When an element (e) has to be deleted from a linked list (h), we need to:
1. Check if linked list is empty. If it is empty, we need not delete anything.
2. If linked list is not empty, we need to traverse through the linked list to find the NODE which contains the element (e). After finding the NODE, we need to change the "next" pointer in the NODE just before the NODE to be deleted to point to the value present in the "next" pointer of the NODE to be deleted.
3. Decrement the "size" variable (as a NODE is deleted) in the linked list HEAD.

```
ListHead DeleteElement(Element e, ListHead h) {
    LINK cur, prev;
    cur=h.start;
    if(cur==NULL)
    {
      printf ("Empty List \n");
      return h;
    }
    while(cur!=NULL)
    {
      if(cur->elem==e)
      {
        if(cur==h.start)
          h.start=cur->next;
        else
          prev->next=cur->next;
        free(cur);
        h.size--;
        break;
      }
      prev=cur;
      cur=cur->next;
    }
    return h;
  }
```

3.3.3 Programming in PERL

173. What will be the output of following Perl code
```perl
my @value_array = ("Index0","Index1");
my $value;
foreach $value (@value_array) {
  $value =~ s/Index//;
}
print "@value_array\n";
```

Answer: 0 1
Value at Array Index will change if we use foreach and substitute scalar "$value"

174. What will be the output of following Perl code
```perl
my @value_array = ("Index0","Index1");
my $value;
for(my $i=0; $i<@value_array; $i++) {
  $value = $value_array[$i];
  $value =~ s/Index//;
}
print "@value_array\n";
```

Answer: Index0 Index1
$value is local to for loop.

175. What is the importance of using -w and "use strict" in perl?

-w is used to flag warnings. It warns about the potential to misinterpret syntax located in the script. A good code ideally should not have any warning.
strict will check the definition and the usage of variables in the script. This is considered a step above the -w command, and can be invoked using the "use strict" command. This stops the execution of the script instead of just giving warnings, when there are any ambiguous or unsafe commands in the script.

176. What will be the output of following Perl code?
```perl
my $line_in_a_file = "I am preparing for an Interview";
my $line_in_a_file =~ s/a/A/;
print "$line_in_a_file\n";
```

First occurrence of "a" will be replaced by "A" and hence the print will display "I Am preparing for an Interview"

177. What will be the output of following Perl code?

```
my $line_in_a_file = "I am preparing for an Interview";
my $line_in_a_file =~ s/a/A/g;
print "$line_in_a_file\n";
```

All occurrences of "a" will be replaced by "A" as we are using "g" (global) in the substitution command. Hence the print will display "I Am prepAring for An Interview"

178. How would you concatenate two strings to form a single string in Perl? Fill in the blank (__?__).

```
my $string1 = "I am preparing ";
my $string2 = "for an Interview";
my $string = __?__
```

A "." concatenates two strings in Perl. Hence, answer for above question will be $string1.$string2;

179. What is the output of following program?

```
#!/usr/bin/perl
use warnings;
use strict;
my $scalar =0;
my @array = ("A","B","C","D");
$scalar = @array;
print "Scalar is $scalar\n";
```

Scalar will store the number of entries in the array. Hence, it will print a value of 4.

180. Mention different special characters that give a special meaning to the "regex" search syntax. Explain use of each.

\ Escape character. Makes the meta-character a literal
^ Specifies start of a string/line
$ Specifies end of a string/line
. Match any character except a newline
* Match zero or more quantifier
+ Match one or more quantifier
- indicates range in a character class (like a-z)
& Substitute complete match
() Grouping characters
[] Character class to match a single character
{} Range quantifiers
<> Anchors that specify left or right word boundary
? Match zero or one quantifier
| Specifies series of alternatives/choices.

181.   List some Quantifiers in regular expressions with their use.

* Match any number of thing(s) it follows (zero or more)
\+ Match one or more thing(s) it follows.
? Match zero or one time the thing it follows.
{N} Match N number of thing(s) it follows.
{N,} Match at-least N number of thing(s) it follows.
{N,M} Match at-least N and at-max M number of thing(s) it follows.

182.   List some Anchors in regular expressions with their use.

^ Match the regular expression from the start of a string/line
$ Match the regular expression at the end of a string/line
< Match the regular expression at the start of a word.
> Match the regular expression at the end of a word.
\b Boundary between a word and a non-word.
\B No Boundary between a word and a non-word.

For Questions 183 to 187: Consider the following code and fill in the blank (__?__) based upon the questions that follow:

```perl
#!/usr/bin/perl
use warnings;
use strict;
my $input_file = "input_file.txt";
my $output_file = "output_file.txt";
my @input_array;

    open(OUTPUT_FILE,'>',$output_file) or die "Cannot Open $output_file file for writing\n$!\n";
    open(INPUT_FILE,'<',$input_file) or die "Cannot Open $input_file for reading\n$!\n";

    while(<INPUT_FILE>){
      if($_ =~ /__?__/){
        print OUTPUT_FILE $_;
      }
    }
    close INPUT_FILE;
    close OUTPUT_FILE;
```

183.   To copy all the lines containing only lowercase alphabets (a to z) in input_file.txt to output_file.txt

Answer: ^([a-z]+)$

184. To copy all the lines containing only lowercase or uppercase alphabets (a to z or A to Z) in input_file.txt to output_file.txt

Answer: ^([a-zA-Z]+)$

185. To copy all the lines containing either lowercase/uppercase alphabets (a to z or A to Z) or digits (0 to 9) in input_file.txt to output_file.txt

Answer: ^([a-zA-Z0-9]+)$

186. To copy all the lines containing $ in them.

Answer: \$

187. To copy all the lines containing only \ or $ in them.

Answer: ^([\\\$]+)$

188. What is the use of the functions chop() and chomp() in Perl?

chop(): chop() function removes the last character of a string and returns that character.
chomp(): chomp() function is an alternative to chop() function. It is most commonly used to remove trailing newline from a string. In general, chomp() function uses Input Record Separator: $/ ($INPUT_RECORD_SEPARATOR) to figure out what to remove from the end of the string.
This function returns the number of the characters removed.

189. What will be the output of the following Perl code?
```
#!/usr/bin/perl
use warnings;
use strict;
my $example_1 = "chop_example";
my $example_2 = "chop_example";
chop($example_1);
my $b = chop($example_2);
print "$example_1 AND $b\n";
```

The print statement will display: `chop_exampl AND e`
chop() removes the last character of the string and returns that character. Therefore, chop($example_1) will remove "e" from $example_1.
chop($example_2) will again remove "e" and since scalar "b" is assigned to chop($example_2), return value ("e" in this case) will be assigned to $b.

190. What will be the output of the following Perl code?

```perl
#!/usr/bin/perl
use warnings;
use strict;
my $example_1 = "chomp_example\n";
my $example_2 = "chomp_example\n";
chomp($example_1);
my $b = chomp($example_2);
print "$example_1 AND $b\n";
```

This will print: `chomp_example AND 1`
chomp() will remove newline character (\n) from both $example_1 and $example_2. It will also return the number of characters removed. Since, scalar "b" is assigned to chomp($example_2), "$b" will contain 1 [as one character (\n) was removed by chomp()]

Chapter 4: Hardware Description Languages

Hardware Description Languages (HDL) are programming languages used to model behavior of digital logic circuits independent of any underlying implementation technology. VHDL and Verilog were the two popular HDLs used for digital logic design and in the recent years SystemVerilog (which is a super set of Verilog) became more widely adopted as it supports object oriented programming concepts and several other features which are very useful for implementing testbenches that are used to verify designs.

This section consists of questions in Verilog and SystemVerilog that cover both the digital logic design modelling as well as testbench modelling concepts.

4.1 Verilog

191. What is the difference between blocking and nonblocking assignments in verilog?

Verilog language supports two types of assignments: blocking and nonblocking. In blocking assignments, evaluation and assignment happens immediately. Thus, if there are multiple blocking assignments in a sequential block, each statement execution follows in a blocking way as shown below.

```
always  @(posedge clk) begin
  x = a|b;
  y = a&b;
  z = x|y;
end
```

In this example, each statement uses blocking assignments and the values of a and b are evaluated and assigned to x and y immediately as the statements execute in order. Hence, in third statement, the new values of x and y are evaluated and assigned to z.

In nonblocking assignments, all assignments are deferred until end of current simulation tick. Hence, evaluation of entire RHS (Right Hand Side) happens first and only then assignment to LHS happens.

```
always  @(posedge clk) begin
  x <= a|b;
  y <= a&b;
  z <= x|y;
end
```

In this example, the RHS of each of the three statements are evaluated first and only after that the assignments to each of LHS (left hand side) happens. Hence, you can notice that in this case, the old values of x and y are OR'ed and assigned to z.

192. **How many flip-flops will be needed when following two codes are synthesized?**

   1)
```
always @(posedge clk) begin
   B = A;
   C = B;
end
```
   2)
```
always @(posedge clk) begin
   B <= A;
   C <= B;
end
```

1) One Flip-flop
2) Two Flip-flops

In first case, blocking assignments are used and hence the value of A will be assigned to B and the new value will be reflected onto C in same cycle and hence the variable B and C results in a wire. So, only one flip flow will be needed.
In second case, old value of B is sampled before the new value is reflected in each cycle. Hence value of A reflects to C only in 2 cycles, resulting in two flip-flops.

193. **What will be the output of a in below code?**
```
always @(posedge clk) begin
   a = 0;
   a <=1;
   $display("a=%0b", a);
end
```

This is because the nonblocking assignment will only reflect at end of cycle, while display will use the current value. Hence, a=0 will be printed.

194. **Write a verilog code to swap contents of two registers (A and B) without any temporary register?**

Using a nonblocking assignment will swap the two values as shown below:
```
always @(posedge clk) begin
   A<=B;
   B<=A;
end
```

195. **What is the output of following code?**
```
module test;
   int alpha,beta;
   initial
```

```
        begin
          alpha = 4;
          beta = 3;
          beta <= beta + alpha;
          alpha <= alpha + beta;
          alpha = alpha - 1;
          $display("Alpha=%0d Beta=%0d", alpha,beta);
        end
      endmodule
```

Note that these assignments are inside an initial begin block without any clocking constructs. The nonblocking assignments will have no effect. Only the blocking assignment of "alpha=alpha-1" will have effect before the display.
Hence answer will be Alpha=3 Beta=3

196. What will be the value of "c" in following 2 cases (after 5 sim units)?
1)
```
    initial begin
      a=0; b=1;
      c = #5 a+b;
    end
```
2)
```
    initial begin
      a=0; b=1;
      #5 c = a+b;
    end
    initial begin
      a=0;
      #3 a=1;
    end
```

1) c=1 after 5 sim units
2) c=2 after 5 sim units

In first case, both a and b are evaluated at time=0, a+b is computed but assignment happens after 5 time units. This is also known as **transport delay.**
In second case, both a and b are evaluated after 5 time units and a+b is assigned to c in the same time unit. Since, a changed to 1 after 3 units (as per second initial block), we get c=2 at end of 5 time units. This is known as **inertial delay.**

197. Analyze following code and find what is wrong with this code that implements a combinational logic?
```
bit a, b, c, d, e;
always @(a, b, c) begin
  e = a & b & c & d;
end
```

A **sensitivity** list is a list of signals that trigger execution of the block when they change values. Since, signal "d" is missing in the sensitivity list, this can cause evaluation of "e" to be not triggered on any changes in "d". This will cause simulation results to look wrong even though synthesis results would be correct.

198. Write a Verilog module for the 3:1 multiplexer that uses the "?:" (conditional operator)

A 3:1 multiplexer has 3 input lines, 2 select lines and an output line which is driven by one of input lines based on select inputs.

```
module mux31_2(inp0,inp1,inp2,sel0,sel1, outres);
   input inp0, inp1, inp2, sel0, sel1;
   output outres;
   assign outres = sel1 ? inp2 : (sel0 ? inp1 : inp0);
endmodule
```

199. What will be the value of X1 and X2 which are modelled using following two always blocks. What is wrong with following coding style?

```
always @(posedge clk or posedge reset)
  if (reset)  X1 = 0; // reset
  else X1 = X2;
always @(posedge clk or posedge reset)
  if (reset) X2 = 1; // set
  else X2 = X1;
```

The Verilog simulators don't guarantee any execution order between multiple always blocks. In above example, since we are using blocking assignments, there can be a race condition and you can see different values of X1 and X2 in multiple simulations. This is a typical example of what a race condition is. If the first always block gets executed before second always block, we will see X1 and X2 to be 1. If the second always block gets executed before first always block, we will see both X1 and X2 to be zero

200. What is the difference between synchronous and asynchronous reset and how do we model synchronous and asynchronous reset using verilog code?

A reset is used to force the state of a design to a known condition after powering up. If a design samples reset on an edge of clock, then it is called as synchronous reset. If the design samples the reset signal without any clock then it is called an asynchronous reset. In terms of implementation, following coding style is used for a synchronous reset

```
always @ (posedge clk) begin
  if (reset) begin
```

```
      ABC <=0;
    end
  end
```
Following coding style is used for an asynchronous reset wherein the reset has highest priority and can happen even without a clock.
```
    always  @ (posedge clk or posedge reset ) begin
      if (reset) begin
        ABC <=0;
      end
    end
```

201. What is the difference between "==" and "===" operators?

Both of these are equality or comparison operators. The "==" tests for logical equality for two states (0 and 1), while the "===" operator tests for logical equality for four states (0, 1, X and Z)

If "==" is used to compare two 4-state variables (logics) and if at least one of them has an X or Z, the result will be X. If the "===" is used to compare two 4-state variables, then comparison is done on all 4 states including X and Z, and the result is 0 or 1.

202. If A and B are two 3-bit vectors initialized as follows:
```
        A = 3'b1x0
        B = 3'b1x0
```
What would be value of following?
1) A==B
2) A===B

1) A==B will only compare non-X/Z values and hence will result in an output "X" if any of the operands has an unknown bit
2) A===B will compare bits including X and Z and hence the comparison would return a 1 as both bit 1 are X for A and B.

203. Write verilog code for a flip-flop and latch and explain differences?

For a flip-flop, the output changes only on the rising or falling edge of a clock signal even if input signal changes in between. However for a latch the output changes as soon as input changes provided the enable signal is high.

Following is the Verilog code for a D flip-flop with synchronous reset.

```
    always @ (posedge clk) begin
      if(reset) begin
        Q <= 0;
        Qbar <= 1;
```

```
    end else begin
        Q <= D;
        Qbar <= ~D;
    end
end
```

Following is the Verilog code for a latch with an enable.

```
always @ (D or Enable) begin
    if(Enable) begin
        Q <= D;
        Qbar <= ~D;
    end
end
```

204. Write Verilog code to detect a pattern of 10110 from an input stream of bits.

Let us assume following states and corresponding meanings:
A: None of the desired pattern is detected yet
B: First bit (1) of the desired pattern is seen
C: First two bits (10) of the desired pattern are seen
D: First three bits (101) of the desired pattern are seen
E: First four bits (1011) of the desired pattern are seen

Based upon the pattern and the states, following will be the state diagram:

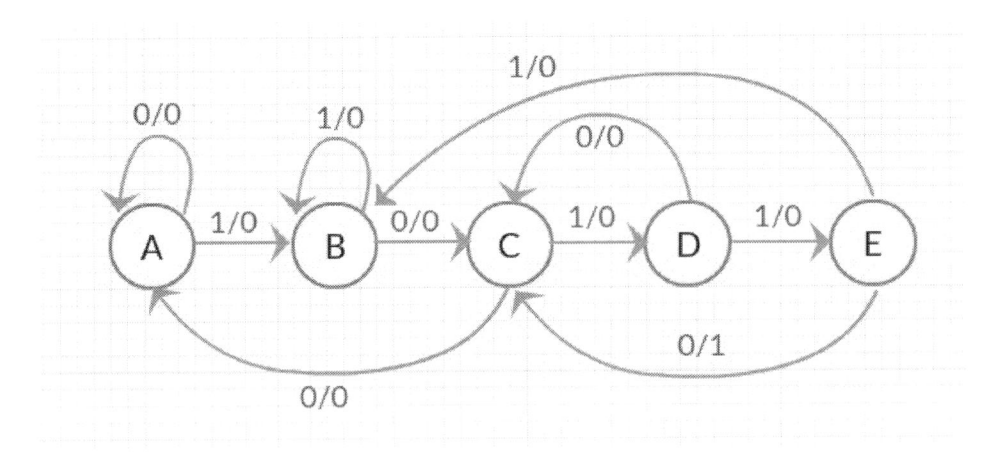

```
module seq_detector(z,x,clock,reset);
    output z;
    input x,clock;
    input reset; //active high
    reg [2:0] state,nextstate;
```

```verilog
    parameter s0=3'b000,s1=3'b001,s2=3'b010,s3=3'b011,s4=3'b100;

  always @ (posedge clock) begin
    if(reset) begin
      state <=s0;
      nextstate<=s0;
    end else begin
      state<=nextstate;
    end
  end

 always @ (x or state)
   case(state)
     s0:   if(x) nextstate=s1; else nextstate=s0;
     s1:   if(x) nextstate=s1; else nextstate=s2;
     s2:   if(x) nextstate=s3; else nextstate=s0;
     s3:   if(x) nextstate=s4; else nextstate=s2;
     s4:   if(x) nextstate=s1; else nextstate=s2;
   endcase
 always @ (x or state)
   case(state)
     s4: if(x) z=1'b0; else z=1'b1;
     s0,s1,s2,s3: z=1'b0;
   endcase
endmodule
```

205. Write Verilog code to print nth Fibonacci number where user provides a value for n. Assume n>2.

The Fibonacci sequence is a series of numbers where a number is found by adding-up the two numbers before it. Starting with 0 and 1, the sequence goes 0, 1, 1, 2, 3, 5, 8, 13, 21, 34, and so forth. Written as a rule, the expression is $x_n = x_{n-1} + x_{n-2}$.
Assuming a max value of n=256, following code will generate the nth fibonacci number. The value of "n" is passed as an input to the module (nth_number)

```verilog
module fibonacci(input clock, reset, input [7:0] nth_number, output [19:0] fibonacci_number);
  reg [19:0] previous_value, current_value;
  reg [7:0] internal_counter;
  reg number_ready;

  always @(posedge reset)
  begin
    previous_value <='d0; //1st Fibonacci Number
    current_value <='d1; //2nd Fibonacci Number
    internal_counter <='d1;
```

```
    end

    always @(posedge clock)
    begin
      internal_counter <= internal_counter + 1;
      current_value <= current_value + previous_value;
      previous_value <= current_value;
      if (internal_counter == (nth_number-2))
        number_ready <= 1;
      else
        number_ready <= 0;
    end

    assign fibonacci_number = current_value;
    always @(number_ready)
      if(number_ready)
        $display("N =%d, Nth Fibonacci Number = %d", nth_number,
fibonacci_number);

endmodule
```

206. Write a Verilog code for Full adder using Half Adder modules.

A full added can be implemented using two instances of half adder modules as shown below:

```
module half_adder(input_0, input_1, sum, carry);
  input input_0, input_1;
  output sum, carry;
  assign sum = (input_0)^(input_1);
  assign carry = (input_0)&(input_1);
endmodule

module full_adder(input_0, input_1, input_2, sum, carry);
  input input_0, input_1, input_2;
  output sum, carry;
  reg sum_intermediate, carry_intermediate_0, carry_intermediate_1;
  half_adder ha1(input0,input1,sum_intermediate,carry_intermediate_0);
  half_adder ha2(sum_intermediate,input2,sum,carry_intermediate_1);
  assign carry = (carry_intermediate_0)|(carry_intermediate_1);
endmodule
```

207. What is the difference between a task and a function in verilog?

1) A function is a subroutine that has to execute without consuming any time or delay while a task is a subroutine that can execute with delays.
2) Functions hence can call other functions but not tasks inside. Tasks can call other tasks as well as functions.
3) Functions could be synthesized while tasks cannot be.
4) Functions normally have a return argument as output but can also have multiple input and reference arguments. Tasks do not return any value but can have multiple input and reference arguments.

4.2 SystemVerilog

208. What is the difference between a reg, wire and logic in SystemVerilog?

reg and wire are two data types that existed from Verilog, while logic is a new data type that was introduced in SystemVerilog.
1) A **wire** is a data type that can model physical wires to connect two elements. Wires can only be driven by continuous assignment statement and cannot hold onto value if not driven. Wires can hence only be used to model combinational logic.
2) A **reg** is a data type that can model a storage element or a state. They need to be driven by an always block and cannot be driven by continuous assignment statement. A reg can be used to model both sequential and combinational logic
3) A **logic** is a new data type in SystemVerilog that can be used to model both wires and state information (reg). It also is a 4 state variable and hence can hold 0, 1, x and z values. If a wire is declared as a logic (wire logic), then it can be used to model multiple drivers and the last assignment will take the value.

209. What is the difference between a bit and logic data type?

bit is a 2-state data type that can take only values 0 and 1, while logic is a 4-state data type which can take values 0, 1, x, and z.
2-state variables will help in a small simulation speed up but should not be used if it is used to drive or sample signals from RTL design in which uninitialized and unknown values will be missed.

210. What is the difference between logic[7:0] and byte variable in SystemVerilog?

byte is a signed variable which means it can only be used to count values till 127. A logic[7:0] variable can be used for an unsigned 8 bit variable that can count up to 255.

211. Which of the array types: dynamic array or associative array, are good to model really large arrays, say: a huge memory array of 32KB?

Associative arrays are better to model large arrays as memory is allocated only when an entry is written into the array. Dynamic arrays on the other hand need memory to be allocated and initialized before using.

For example: If you want a memory array of 32KB to be modelled using dynamic array, you would first need to allocate 32K entries and use the array for read/write. Associative arrays doesn't need allocation and initialization of memory upfront and can be allocated and initialized just when an entry of the 32K array needs to be referenced..

However, associative arrays are also slowest as they internally implement search for elements in the array using a hash.

212. Suppose a dynamic array of integers (myvalues) is initialized to values as shown below. Write a code to find all elements greater than 3 in the array using array locator method "find"?

```
int myvalues [] = '{9,1,8,3,2,4,6},
```

```
int match_q[$];
match_q = myvalues.find with (item > 3);
```

213. What is the difference between a struct and union in SystemVerilog?

A **structure** represents a collection of data types that can be referenced as a whole, or the individual data types that make up the structure can be referenced by name. For example: in the example below, we have a struct defined called `instruction_s` that groups a 24 bit address field and an 8 bit opcode field.

```
typedef struct {
  bit [7:0] opcode;
  bit [23:0] addr;
} instruction_s;
instruction_s  current_instruction;
current_instruction.addr='h100;
```

The `instruction_s` struct can be referenced together or individual members can be accessed. The total memory allocated would be the sum of memory needed for all the data types. Hence in above example, the `currect_instruction` struct would take a total memory of 32 bits (24 bit address and 8 bit opcode)

A **union** is a data type which can be accessed using one and only one of the named member data type. Unlike struct you cannot access all member data types together. The memory allocated for the union would be the maximum of the memory needed for the member data types. Unions are normally useful if you want to model a hardware resource like register that can store values of different types. For example: if a register can store an integer and a real values, you can define a union as follows:

```
typedef  union {
  int  data;
```

```
    real  f_data;
} state_u;
state_u  reg_state;
reg_state.f_data = 'hFFFF_FFFF_FFFF_FFFF;
$display(" int_data =%h", reg_state.data);
```

In this example, the union state_u can either hold a 32 bit integer data or it can hold 64 bit real data. Hence, the memory allocated for the union reg_state will be 64 bits (bigger of the two data types). Since, there is shared memory for all member data types, in above example, if we assign a 64 bit value to reg_state.f_data, we will be also able to reference the 32 bit of same using the other data type.

214. What is the concept of a "ref" and "const ref" argument in SystemVerilog function or task?

A **ref** keyword is used to pass arguments by reference to a function instead of a value. The subroutine/function shares the reference handle with the caller to access values. This is an efficient way of passing arguments like class objects or arrays of objects where otherwise creating a copy would consume more memory on the stack. Also, since the caller and the function/task shares same reference, any change done inside the function using the reference would also be visible to the caller.

For Example: Here is an example of a CRC function which needs a big packet as argument to compute CRC. By passing as reference, each call to CRC function doesn't need to create a copy of the packet on stack memory.

```
function automatic int crc(ref byte packet [1000:1] );
  for( int j= 1; j <= 1000; j++ ) begin
    crc ^= packet[j];
  end
endfunction
```

A **const** keyword is used if user wants to make sure that the ref argument is not modified by the function. For example: in the same CRC function, the argument can be declared as a "const ref" argument as shown below to make sure that the original packet contents are not modified accidentally by the CRC function.

```
function automatic int crc( const ref byte packet [1000:1] );
  for( int j= 1; j <= 1000; j++ ) begin
    crc ^= packet[j];
  end
endfunction
```

215. What would be the direction of arguments a and b in following?
```
task sticky(ref int array[50], int a, b);
```

Each argument of a task or function can have a direction which can be one of input, output, inout or ref. If no direction is specified, the default value of input is selected. If one of the arguments specifies a direction, then all following arguments hold on to same direction unless explicitly changed.

So in above example, the first argument has a direction defined as "ref" which means it is an argument passed by reference. Since the following arguments "a" and "b" have no explicit directions defined, these also get the same direction. Hence, "a" and "b" also become pass by reference arguments.

216. What is the difference between a packed array and an unpacked array?

A packed array represents a contiguous set of bits while an unpacked array need not be represented as a contiguous set of bits. In terms of difference in declarations, following is how a packed and unpacked array is declared

```
bit [7:0] data ; // packed array of scalar bit types
real latency [7:0]; // unpacked array of real types
```

Packed arrays can be made of only the single bit data types (bit, logic, reg), or enumerated types. Example: `logic[31:0] addr; //packed array of logic type`

Unpacked arrays can be made of any data type. Example:
```
class  record_c;
record_c  table[7:0];   //unpacked array of record objects
```

217. What is the difference between a packed and unpacked struct?

A packed structure is a way in which a packed bit vector can be accessed as struct members. Or in other words, if all the members of a struct consist of only bit fields and can be packed in memory without any gaps, it can be a packed structure. For example: in the structure definition below, all the members can be represented as bit vectors (int is equivalent to 32 bits, short int to 16 bits, byte to 8 bits) and a struct can be packed into a single contiguous memory of 56 bits.

```
struct packed {
  int a;
  short int b;
  byte c;
} pack1_s;
```

An unpacked struct need not be packed into contiguous set of bits and hence different members could be placed in memory with gaps. Following is an example with a structure having different data types that cannot be packed in memory.

```
struct  record {
  string  name;
  int  age;
  string parent;
} record_s
```

218. Which of the following statement is true?
1) Functions should execute in Zero Simulation Time.
2) Tasks should execute in Zero Simulation Time.

1) True
2) False

Functions always need to be executed in zero simulation time and cannot contain any construct that can induce a time delay (Example: waiting for clock edge or # delays etc.). Tasks can have constructs causing timing delays and hence, need not complete execution in zero time.

219. Given a dynamic array of size 100, how can the array be re-sized to hold 200 elements while the lower 100 elements are preserved as original?

A dynamic array needs memory allocation using new[] to hold elements. Here is an example with an integer array that grows from an initial size of 100 elements to 200 elements.

```
integer addr[]; // Declare the dynamic array.
addr = new[100]; // Create a 100-element array.
.........
// Double the array size, preserving previous values.
addr = new[200](addr);
```

220. What is the difference between "forever" and "for" in SystemVerilog ?

The "forever" loop repeatedly executes a statement without any limit. The only way execution can stop is by using a break statement. A forever loop if used without any timing controls (like clock or time delay) can result in a zero-delay infinite loop and cause hang in simulation.

The "for" loop is used for executing a statement for a defined number of times based on conditions that are defined.

221. What is the difference between "case", "casex" and "casez" in SystemVerilog?

The case statement is a multiway decision statement that tests whether an expression matches one of a number of possible values and branches accordingly. Following is a simple example to implement a 3:1 MUX using a case statement

```
case (select[1:0])
  2'b00:  out_sig = in0;
  2'b01:  out_sig = in1;
  2'b10:  out_sig = in2;
  default: out_sig = 'x
```

```
endcase
```

In the above example of using a "case" statement, the expression match happens exactly with what is specified. For example, in above case statement, if at least one of the select lines is X or Z, then it will not match any conditions and will execute the default statement.

"casez" is a special version of case expression which allows don't cares in comparison of the expressions. These are typically useful in decoder logic which only treats fewer bits. Here is an example where a 3 bit interrupt request queue (irq) bus is decoded into 3 separate interrupt pins based on which bit in the bus is high, while other bits are don't care.

```
casez (irq)
  3'b1?? : int2 = 1'b1;
  3'b?1? : int1 = 1'b1;
  3'b??1 : int0 = 1'b1;
endcase
```

"casex" is another special version where in addition to don't cares, it also ignores X and Z values in comparison.

222. Which of the logical equality operators "==" or "===" are used in case expression conditions for case, casex, and casez?

All of the 3 case statements use "===" logical equality comparison to evaluate condition matches.

223. What is the difference between $display, $write, $monitor and $strobe in SystemVerilog?

1) `$display` : Print the values immediately when executed.
2) `$strobe` : Print the values at the end of the current timestep.
3) `$monitor` : Print the values at the end of the current timestep if any values change. If $monitor is called more than once, the last call will override previous one.
4) `$write` : This is same as $display but doesn't terminate with a newline (\n).

224. What is wrong with following SystemVerilog code?

```
task wait_packet;
  Packet packet;
  event packet_received;
  @packet_received;
  packet = new();
endtask
function void do_print();
  wait_packet();
```

```
        $display("packet received")
     endfunction
```

A function cannot have any construct that consumes time. In above example, the function do_print() is calling a task which consumes time. Hence, this is illegal.
A proper fix is to have the function do_print() be called inside the task to print packet after it is received.

225. What is the difference between new() and new[ ] in SystemVerilog?

The function **new()** is the class constructor function in SystemVerilog. It is defined in a class to initialize data members of the class.
The **new[ ]** operator is used to allocate memory for a dynamic array. The size of the dynamic array that needs to be created is passed as an argument to the **new[ ]**.

226. What is the concept of forward declaration of a class in SystemVerilog?

Sometimes a class might reference another class which is not fully defined in the compile order. This can cause a compile error. For Example: If two classes Statistics and Packet are defined in following order, then while compiling Statistics class, the definition of packet is not yet seen and compiler will fail.

```
    class Statistics;
       Packet p1;
    endclass

    class Packet;
       //full definition here
    endclass
```

To avoid this problem, the Packet Class can be forward declared before the full definition. This concept is called forward declaration.

```
    typedef Packet;   //forward declaration
    class Statistics;
       Packet p1;
    endclass

    class Packet;
       //full definition here
    endclass
```

227. Analyze following code and explain if there are any issues with code?

```
    task gen_packet(Packet pkt);
      pkt = new();
      pkt.dest = 0xABCD;
```

```
    endtask

    Packet pkt;
    initial begin
      gen_packet(pkt);
      $display(pkt.dest);
    end
```

The code will result in a runtime null pointer error.
The task `gen_packet()` has an argument which is pass by value. It creates an object and uses the handle of the argument.
In the initial block, once the `gen_packet()` task is called and once the `pkt.dest` field is modified it is still local to the task and trying to display the value outside the task causes a null pointer error.

228. What is the difference between private, public and protected data members of a SystemVerilog class?

1) **Private** data members of a class can only be accessed from within the class. These data members will not be visible in derived classes.
2) **Public** members can be accessed from within the class as well as outside the class also. These are also visible in derived classes.
3) **Protected** data members are similar to private members in the sense that they are only accessible within the class. However, unlike private members, these are also visible in derived classes.

229. Are SystemVerilog class members public or private by default ?

SystemVerilog class members are public by-default, unlike other languages like C++/Java which have default data members as private.

230. What is a nested class and when would you use a nested class?

When the definition of a class contains another class definition, then that class is called a nested class. For example: In the code below, the StringList class definition contains definition for another class Node.

```
    class StringList;
      class Node; // Nested class for a node in a linked list.
        string name;
        Node link;
      endclass
    endclass
```

Nesting allows hiding of local names and local allocation of resources. This is useful when a new type is needed as part of the implementation of a class.

231. What are interfaces in SystemVerilog?

The interface construct in SystemVerilog is a named bundle of nets of variables which helps in encapsulating communication between multiple design blocks. An interface can be instantiated in a design and can be connected using a single name instead of having all the port names and connections.

In addition to connectivity, functionality can also be abstracted in an interface as it supports defining functions that can be called by instantiating design for communication. Interfaces also support procedural (always/initial blocks) and continuous assignments which are useful for verification in terms of adding protocol checks and assertions.

Following is a simple example on how an interface can be defined.

```
interface simple_bus; // Define the interface
   logic req, gnt;
   logic [7:0] addr, data;
   logic [1:0] mode;
   logic start, rdy;
endinterface: simple_bus
```

232. What is a modport construct in an interface?

modport (short form for module port) is a construct in an interface that let you group signals and specify directions. Following is an example of how an interface can be further grouped using modports for connecting to different components.

```
interface arb_if(input bit clk);
   logic [1:0] grant, request;
   logic reset;
   modport TEST (output request, reset, input grant, clk);
   modport DUT (input  request, reset, clk, output grant);
   modport MONITOR (input request, grant, reset, clk);
endinterface
```

In this example, you can see that the same signals are given different directions in different modports. A monitor component needs all signals as input and hence the modport MONITOR of interface can be used to connect to monitor. A test or a driver will need to drive some signals and sample other signals and above example shows a modport TEST that can be used

233. Are interfaces synthesizable?

Yes, interfaces are synthesizable.

234. What is a clocking block and what are the benefits of using clocking blocks inside an interface?

A **clocking block** is a construct that assembles all the signals that are sampled or synchronized by a common clock and define their timing behaviors with respect to the clock. Following example illustrates a simple clocking block.

```
clocking sample_cb @(posedge clk);
  default input #2ns output #3ns;
  input a1, a2;
  output b1;
endclocking
```

In above example, we have defined a clocking block with name `sample_cb` and the clock associated with this clocking block is clk. The *default* keyword defines the default skew for inputs (2 ns) and output (3 ns). The input skew defines how many time units before the clock event the signal is sampled. The output skew defines how many time units after the clock event the signal will be driven.

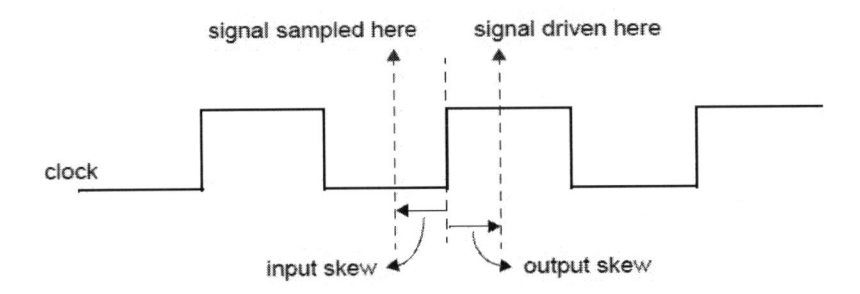

A clocking block can be declared only inside a module or an interface.

235. What is the difference between following two ways of specifying skews in a clocking block?
1) input   #1 step req1;
2) input   #1ns   req1;

The clocking skew determines how many time units away from the clock event a signal is to be sampled (input skew) or driven (output skew). A skew can be specified in two forms - either explicitly in terms of time as in case 2) above, where the signal is sampled **1ns** before the clock, OR in terms of time step as in case 1) above, where the `step` corresponds to global time precision (defined using `` `timescale `` directive)

236. What are the main regions inside a SystemVerilog simulation time step?

A SystemVerilog simulator is an event driven simulator and as the simulator advances in time, it needs to have a well-defined manner in which all events are scheduled and executed. In any event simulation, all the scheduled events at a specific time defines a time slot. A time slot is divided into a set of ordered regions to provide predictable interactions between the design and testbench code.

A timeslot can be broadly divided into 5 major regions as shown below and each of the regions can be further subdivided into sub-regions.

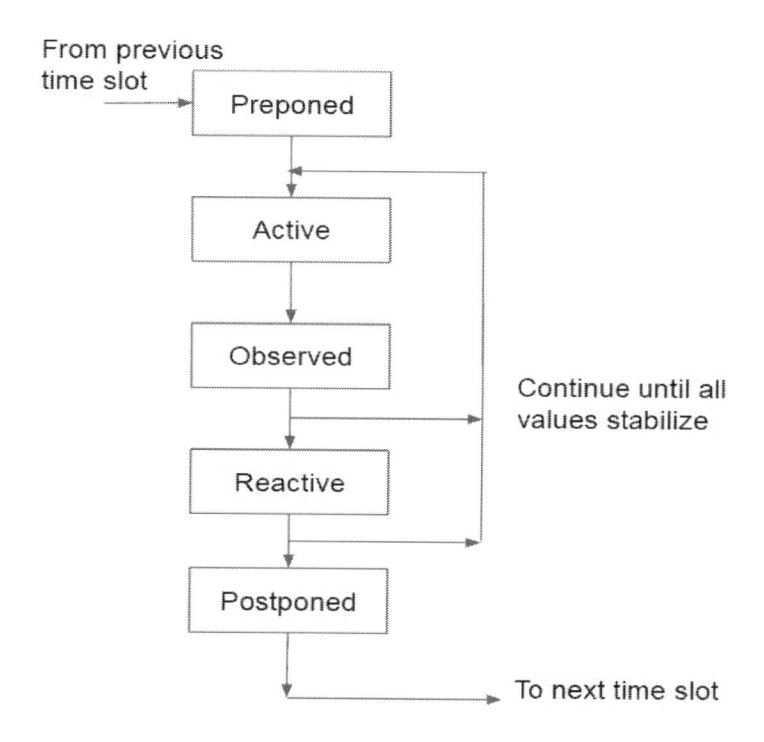

```
From previous
time slot
    ──────▶ ┌───────────┐
            │ Preponed  │
            └───────────┘
                  │
                  ▼
            ┌───────────┐ ◀──────┐
            │  Active   │        │
            └───────────┘        │
                  │              │
                  ▼              │
            ┌───────────┐        │
            │ Observed  │        │
            └───────────┘        │   Continue until all
                  │      ────────▶   values stabilize
                  ▼              │
            ┌───────────┐        │
            │ Reactive  │        │
            └───────────┘        │
                  │      ────────┘
                  ▼
            ┌───────────┐
            │ Postponed │
            └───────────┘
                  │
                  └──────▶ To next time slot
```

1) **Prepone**: The preponed region is executed only once and is the first phase of current time slot after advancing the simulation time. Sampling of signals from design for testbench input happens in this region.
2) **Active**: The active region set consists of following sub regions - Active, Inactive and the NBA (Nonblocking assignment) regions. RTL code and behavioral code is scheduled in Active region. All blocking assignments are executed in Active region. For nonblocking assignments, evaluation of RHS happens in Active region, while assignment happens in the NBA region. If there are any assignments with #0 delays, those happen in the Inactive region.

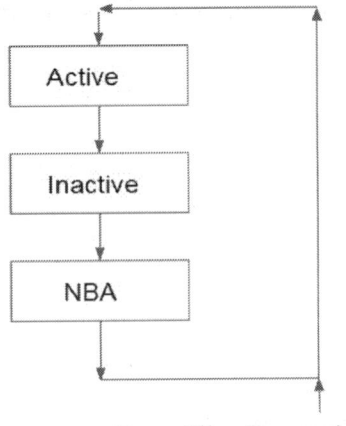

```
                    ┌────────┐
                    │ Active │
                    └────────┘
                         │
                    ┌────────┐
                    │Inactive│
                    └────────┘
                         │
                    ┌────────┐
                    │  NBA   │
                    └────────┘
                         │
              From Reactive regions
```

3) **Observed:** The Observed region is for evaluation of property expressions (used in concurrent assertions) when they are triggered. During property evaluation, pass/fail code is scheduled for later in the Reactive region of the current time slot
4) **Reactive**: The reactive region set (Re-active, Re-Inactive and Re-NBA) is used to schedule blocking assignments, #0 blocking assignments and nonblocking assignments included in SystemVerilog "program" blocks. This separate Reactive region ensures that all the design code evaluation (in Active region set) stabilizes before the testbench code in the program blocks is evaluated. With OVM/UVM methodologies, there is no need of program blocks (with standard phasing of testbench code) and hence Reactive region may not be used much.
5) **Postponed:** There is also a postponed region which is the last phase of current time slot. $monitor, $strobe and other similar events are scheduled for execution in this region. $display events are scheduled for execution in Active and Reactive regions (if called in program blocks).

237. **Given following constraints, which of the following options are wrong?**
```
rand logic [15:0] a, b, c;
constraint c_abc {
  a < c;
  b == a;
  c < 30;
  b > 25;
}
```
1) b can be any value between 26 and 29
2) c can be any value between 0 and 29
3) c can be any value between 26 and 29

SystemVerilog constraints are bidirectional. In above example, since a <c and c < 30, a has to be less than 30. However, since b also has to be equal to a and > 25, it means a, b, and c can only take values from 26 to 29 only. Hence, only option 2) is wrong.

238. Will there be any difference in the values generated in following constraints?

1)
```
class ABSolveBefore;
  rand bit A;
  rand bit [1:0] B;
  constraint c_ab {
    (A==0) -> B==0;
    solve A before B;
  }
endclass
```

2)
```
class ABSolveBefore;
  rand bit A;
  rand bit [1:0] B;
  constraint c_ab {
    (A==0) -> B==0;
    solve B before A;
  }
endclass
```

In both the cases, by default, A can have a value of 0 or 1, while B can have a value of 0,1,2,3. However, since there is a constraint that if A==0 -> B==0, this restricts value of B to be zero if A is zero.

1) If we solve A first, then A=0 or 1 is picked with ½ probability each. If A is picked 0, B will be always zero. If A is picked one, then B has equal probability of taking 3 values (1,2,3).
2) If we solve B first, then B = 0,1,2,3 has equal probability of ¼. If B is picked 0, then A will be zero. If B is non-zero, then A will be always 1.

So, in both cases, the values generated for A and B will be same, but probability of generation of values will differ based on which is solved first.

239. What is a unique constraint in SystemVerilog?

A unique constraint is used to randomize a group of variables such that no two members of the group have the same value. Following shows an example: Here a class has a random array of bytes (a) and one another byte (b). The unique constraint in this example shows how unique values can be generated for all of these.

```
class Test;
  rand byte a[5];
  rand byte b;
  constraint ab_cons { unique {b, a[0:5]}; }
endclass
```

240. How can we disable or enable constraints selectively in a class?

<object>.constraint_mode(0) :: To disable all constraints
<object>.<constraint>.constraint_mode(0) :: To selectively disable specific constraints

```
    class ABC;
      rand int length;
      rand byte SA;
      constraint c_length { length inside [1:64];}
      constraint c_sa {SA inside [1:16];}
    endclass
    ABC abc = new();
    abc.constraint_mode(0);// will turn off all constraints
    abc.c_length.constraint_mode(0);// will turn off only length
constraint
```

241. Given a Packet class with following constraints, how can we generate a packet object with address value greater than 200?

```
    class Packet;
      rand bit[31:0] addr;
      constraint c_addr { addr inside [0:100];}
    endclass
```

Since default constraint restricts address to be less than 100, we will need to use inline constraints and turn off default constraint as below:

```
    Packet p = new();
    p.c_addr.constraint_mode(0);
    p.randomize() with {addr > 200;};
```

242. What are pre_randomize() and post_randomize() functions?

These are built-in callback functions supported in SystemVerilog language to perform an action immediately either before every randomize call or immediately after randomize call. A pre_randomize() is useful for setting or overriding any constraints while a post_randomize() is useful to override results of a randomization.

243. Write constraints to generate elements of a dynamic array (abc as shown in code below) such that each element of the array is less than 10 and the array size is less than 10.

```
    class dynamic_array;
      rand unsigned int abc[];
    endclass
```

For dynamic arrays, we can use a foreach constraint to constraint the value of each of the element of the array as shown below:
```
constraint c_abc_len {
  abc.size() < 10;
  foreach (abc[i])
    abc[i] < 10;
}
```

244. Write constraints to create a random array of integers such that array size is between 10 and 16 and the values of array are in descending order?
```
class array_abc;
   rand unsigned int  myarray[];
endclass
```

```
constraint  c_abc_val {
  myarray.size inside { [10:16] };
  foreach (myarray[i])
    if (i>0) myarray[i] < myarray[i-1];
}
```

245. How can we use constraints to generate a dynamic array with random but unique values ? Refer the code below:
```
class TestClass;
   rand bit[3:0] my_array[]; //dynamic array of bit[3:0]
endclass
```

There are two ways in which this can be done - one using the SV unique constraint and one without using it as shown in 2) below.

1) Add a unique constraint to the class as below
```
constraint c_rand_array_uniq {
  my_array.size == 6;   //or any size constraint
  unique {my_array};    //unique array values
}
```

2) Without using unique constraint, you can still generate incremental values and then do an array shuffle() in post_randomize();
```
constraint c_rand_array_inc {
  my_array.size == 6 ;// or any size constraint
  foreach (my_array[i])
    if(i >0)  my_array[i] > my_array[i-1];
}
function post_randomize();
```

```
        my_array.shuffle();
    endfunction
```

246. Given a 32 bit address field as a class member, write a constraint to generate a random value such that it always has 10 bits as 1 and no two bits next to each other should be 1

```
class packet;
  rand bit[31:0] addr;
  constraint c_addr {
    $countones(addr) ==10;
    foreach (addr[i])
      if(addr[i] && i>0)
        addr[i] != addr[i-1];
  }
endclass
```

247. What is the difference between "fork - join", "fork - join_any" and "fork - join_none"?

SystemVerilog supports three types of dynamic processes that can be created at run-time and executed as independent threads from the processes that spawned them.

1) **fork .. join:** Processes that are created using "fork .. join" run as separate threads but the parent process that spawned them stall until a point where all threads join back together. If we look at the example below: there are three processes - task1, task2 and task3, that will run in-parallel and only after all three of these complete, the $display() after the join statement will execute.
    ```
    initial begin
      fork
        task1; // Process 1
        task2; // Process 2
        task3; // Process 3
      join
      $display("All tasks finished");
    end
    ```

2) **fork .. join_any:** Processes that are created using "fork … join_any" run as separate threads but the parent process that spawned those stalls only until any one of the threads complete. Further, the remaining threads and the parent process can run parallely. If we look at the example below: there are three processes - task1, task2 and task3 that will run parallely. When one of task1/task2/task3 completes, the join_any will complete and cause the $display() to execute while other tasks might still be running.
    ```
    initial begin
    ```

```
    fork
      task1; // Process 1
      task2; // Process 2
      task3; // Process 3
    join_any
      $display("Any one of task1/2/3 finished");
  end
```

3) **fork .. join_none:** Processes that are created using "fork ... join_none" run as separate threads but the parent process that spawned them doesn't stall and also proceed parallely. Refer to the following example and there are three processes - task1, task2 and task3 that will run parallely with the parent process. .

```
  initial begin
    fork
      task1; // Process 1
      task2; // Process 2
      task3; // Process 3
    join_none
      $display("All tasks launched and running");
  end
```

248.   What is the use of "wait fork" and "disable fork" constructs?

When using a "fork..join_none" or a "fork..join_any", sometimes we will want to synchronize the parent process with the dynamic threads running parallely and this can be done using **wait fork** construct as follows:

```
  initial begin
    fork
      task1; // Process 1
      task2; // Process 2
    join_none
    $display("All tasks launched and running");
    wait fork;
    $display("All sub-tasks finished now");
  end
```

Similarly, a **disable fork** can be used to prematurely stop the child forked dynamic processes as shown below:

```
  initial begin
    fork
      task1; // Process 1
      task2; // Process 2
    join_any
    $display("One of task1/2 completed ");
    disable fork;
    $display("All other tasks disable now");
```

end

249. What is the difference between hard and soft constraints?

The normal constraints that are written in SystemVerilog classes are known as **hard constraints**, and the constraint solver need to always solve them or result in a failure if it cannot be solved.
On the other hand, if a constraint is defined as **soft**, then the solver will try to satisfy it unless contradicted by another hard constraint or another soft constraint with a higher priority.
Soft constraints are generally used to specify default values and distributions for random variables and can be overridden by specialized constraints.

```
class Packet;
  rand int length;
  constraint length_default_c { soft length inside {32,1024}; }
endclass

Packet p = new();
p.randomize() with { length == 1512; }
```

In the above example, if the default constraint was not defined as soft, then the call to randomize would have failed.

250. What will be the value of result printed from each of the threads in below code?

```
initial begin
  for (int j=0; j<3; j++) begin
    fork
      automatic int result;
      begin
        result= j*j;
        $display("Thread=%0d result=%0d", j, result);
      end
    join_none
    wait fork;
  end
end
```

Since "j" is not an automatic variable per thread, it keeps incrementing after each thread is spawned, and when all threads start executing each of them will see a value of 3. Hence, each thread will print out 9 as a result. If each thread is intended to use different values, we should copy the value of "j" to an automatic variable as below:

```
automatic int  k = j;
begin
  result = k*k;
```

end

251. How many parallel processes does this code generate?
```
fork
    for (int i=0; i < 10; i++ ) begin
        ABC();
    end
join
```

Since the "for" loop is inside the fork join, it executes as a single thread.

252. What is wrong with following SystemVerilog constraint?
```
class packet;
    rand bit [15:0] a, b, c;
    constraint  pkt_c { 0 < a < b < c; }
endclass
```

There can be a maximum of only one relational operator (<, <=, ==, >=, or >) in an expression. If multiple variables need to be in some order, we will need to write multiple expressions as below.
```
constraint  pkt_c { 0 < a;   a < b ;   b < c; }
```


253. Which keyword in SystemVerilog is used to define Abstract classes?

Sometimes a class is defined with an intention to be only a base class from which other classes can be derived. Methods may also be defined as virtual in an abstract class without any definition. Such a base class with no intention to create an object is defined as an abstract class and is identified using the **virtual** keyword as prefix in the class definition.
For Example:
```
virtual class  A;
    virtual function process();
endclass: A
```


254. What is the difference between a virtual function and a pure virtual function in SystemVerilog?

A function in a class is defined as virtual to allow overriding the implementation of the function in a derived class. The base class may or may not have an implementation of the function that is virtual and may or may not be overridden in the derived class.

A **pure virtual function** is a kind of virtual function which will have only declaration without any implementation. Any class that derives from a base class having "**pure virtual**" functions need to implement the function. Pure virtual functions are normally used in abstract class definitions. See the following example of usage of same.

```
virtual class BasePacket;
  // No implementation
  pure virtual function integer send(bit[31:0] data);
endclass

class EtherPacket extends BasePacket;
  virtual function integer send(bit[31:0] data);
    // body of the function
    // that implements the send
        ….…
  endfunction
endclass
```

255. What does keyword "extends" represent in SystemVerilog?

The "extends" keyword is used in class definition to specify the class from which the class is derived. For Example: `class A extends B;` means class A is derived from class B.

256. What are Semaphores? When are they used?

Semaphore is a mechanism used to control access to shared resources. It can be considered like a bucket with a number of keys created when the semaphore is created. Multiple processes that use a semaphore to access a shared resource should first procure a key from the bucket before they can continue to execute. This guarantees that processes which do not get a key will wait until the ones that procured keys releases them back. Semaphores are typically used for mutual exclusion, access control to shared resources, and basic synchronization. Following is how a semaphore can be created.

```
semaphore smTx;
smTx = new(1);  //create the semaphore with 1 keys.
```

The methods `get()` (blocking call) and `try_get()` (nonblocking call) are used to get keys from semaphore while `put()` method is used to release keys back.

257. What are Mailboxes? What are the uses of a Mailbox?

A mailbox is a communication mechanism that allows messages to be exchanged between processes. Data can be sent to a mailbox by one process and retrieved by another. Following is an example of declaring and creating a mailbox:

```
mailbox mbxRcv;
mbxRcv = new();
```

To place a message in a mailbox, two methods are supported `put()` (blocking) and `try_put()` (nonblocking). To retrieve a message from mailbox, two methods are supported `get()` (blocking) and `try_get()` (nonblocking). To retrieve the number of messages in the mailbox, we can use `num()`.

258. What is difference between bounded and unbounded mailboxes? How can we create unbounded mailboxes?

A mailbox is called **bounded** if the size of mailbox is limited when created.
```
mailbox mbxRcv;
mbxRcv = new(10);   //size bounded to 10
```

A mailbox is **unbounded** if the size is not limited when created.
```
mailbox mbxRcv;
mbxRcv = new();    //size is unbounded or infinite
```

A bounded mailbox becomes full when it contains the bounded number of messages and any further attempt to place a message will cause the process to be suspended while unbounded mailboxes never suspend a thread in a send operation.

259. What is an "event" in SystemVerilog? How do we trigger an "event" in SystemVerilog?

An identifier declared as an event data type is called a named event. Named event is a data type which has no storage. A named event can be triggered explicitly using "->" operator. A process can use the event control "@" operator to block execution until the event is triggered. Events and event control gives a powerful and efficient means of synchronization between two or more concurrently running process.

Following example pseudo code shows how two processes can synchronize execution using an event. The `send_req()` task emits an event once a request is send while the `receive_response()` event waits until `req_send` event is seen

```
module test;
  event req_send;

  initial begin
    fork
      send_req();
      receive_response);
    join
  end

  task send_req();
    //create and send a req
```

```
    -> req_send; //trigger event
  endtask

  task receive_response();
    @req_send; //wait until a send event is triggered
    //collect response
  endtask
endmodule
```

260. How can we merge two events in SystemVerilog?

An event variable can be assigned to another event variable. When an event variable is assigned to other, both the events point to same synchronization object and are said to be merged.

261. What is std::randomize() method in SystemVerilog and where is it useful?

The *std::randomize()* is a scope randomize function that enables users to randomize data in the current scope without the need to define a class or instantiate a class object. This is useful if some variables required to be randomized are not part of a class. Refer following example of a function inside a module. There are few variables inside a module that can be randomized using std::randomize().

```
module stim;
  bit [15:0] addr;
  bit [31:0] data;

  function bit gen_stim();
    bit success, rd_wr;
    success = std::randomize(addr, data, rd_wr);
    return rd_wr ;
  endfunction
  ...
endmodule
```

std::randomize() behaves similar to randomize() function of classes and can take all kinds of constraints supported in a class. For example if we want to add a constraint, it can be added using the **with** construct as follows:

```
success = std::randomize( addr, data, rd_wr ) with {rd_wr -> addr > 'hFF00;};
```

262. Is it possible to override a constraint defined in the base class in a derived class and if so how?

Yes, a constraint defined in the base class can be overridden in a derived class by changing the definition using the same constraint name. For Example: Refer the constraint c_a_b_const in following code. In the base class, it is defined to always have a value of a < b, but in a derived class, it has been overridden to have always a > b.

```
class Base;
  rand int a ;
  rand int b;
  constraint  c_a_b_const {
    a < b;
  }
endclass

class  Derived extends Base;
  constraint c_a_b_const {
    a > b;
  }
endclass
```

263. Identify what could be wrong if following function is called in SystemVerilog constraint as below?

```
function int count_ones ( ref bit [9:0] vec );
  for( count_ones = 0; vec != 0; vec = vec >> 1 ) begin
    count_ones += vec & 1'b1;
  end
endfunction

constraint C1 { length == count_ones( vec ) ; }
```

You cannot have functions with arguments as reference in constraints. "const ref" is fine which will not allow the values to be changed inside function

264. Is there any difference between following two derived class codes?
```
virtual class Base;
  virtual function printA();
  endfunction
endclass
```
1)
```
class Derived extends Base;
  function printA();
```

131

```
            //new print implementation
         endfunction
      endclass
2)
      class Derived extends Base;
         virtual function printA();
            //new print implementation
         endfunction
      endclass
```

There will be no difference. Using virtual keyword to override a default function is not necessary in derived class, but neither there is an error.

265. **Find issues (if any) in following code?**
```
class Packet;
   bit [31:0] addr;
   bit err = 0;
endclass
class ErrPacket extends Packet;
   bit err = 1;
endclass

module Test;
   initial begin
      Packet p;
      ErrPacket  ep;
      ep = new();
      p = ep;
      $display("packet addr=%h err=%b", p.addr, p.err);
   end
endmodule
```

No issues. A base class handle can be used to reference a derived class object, while the other way (derived class handle referencing base class object will not work)

266. **Given following definitions of two classes - Packet and Bad Packet, find the correct sequence of which of the `compute_crc()` functions gets called in example code?**

```
class Packet; //Base Class
    rand bit [31:0] src, dst, data[8]; // Variables
    bit [31:0] crc;
    virtual function void compute_crc;
        crc = src ^ dst ^ data.xor;
    endfunction
endclass : Packet


class BadPacket extends Packet;   //Derived class
    rand bit bad_crc;
    virtual function void compute_crc_crc;  //overriding definition
        super.compute_crc();   // Compute good CRC
        if (bad_crc) crc = ~crc; // Corrupt the CRC bits
    endfunction
endclass : BadPacket
```

Example Code (Usage): Answers 1), 2), and 3) are mentioned in comments

```
Packet pkt;
BadPacket   badPkt;
initial begin
    pkt  = new;
    pkt.compute_crc; // 1) Which of compute_crc() gets called?
    badPkt = new;
    badPkt.compute_crc; // 2) Which of compute_crc() gets called ?
    pkt = badPkt; // Base handle points to ext obj
    pkt.compute_crc; // 3) Which of compute_crc() gets called ?
end
```

1) Calls base class `Packet::compute_crc()`
2) Calls derived class `BadPacket::compute_crc()`
3) Calls derived class `BadPacket::compute_crc()`, as the base class pointer still references a derived class object.

This is because when virtual methods are used, SystemVerilog uses the type of the object and not the handle to decide which routine to call.

267. **Which of following constraint coding (giving same results) is better in terms of performance and why?**

1)
```
constraint align_addr_c {
  addr%4 == 0;
}
```

2)
```
constraint align_addr_c {
  addr[1:0] == 0;
}
```

Arithmetic operations are more costly. Hence, 2) gives better performance

268. **Analyze following pseudo code and comment on any inefficiency? If so, how can we change the code to make it more efficient**

```
class Packet;
  rand byte[] data;
  rand bit valid;
endclass
Packet lots_of_pkts[$];  //array of packets
size = lots_of_pkts.size();
for (i = 0; i < size; i++) begin
  lots_of_pkts[i].randomize();
  valid = lots_of_pkts[i].valid
  if (valid == TRUE) begin
    inject(lots_of_pkts[i].data);
  end
end
```

Every time the packet class is randomized, the member valid can be either 0 or 1. Calling `randomize()` and then selectively deciding to use the results is not an efficient way as randomize is compute heavy. To improve, we should `randomize()` using an inline constraint to have bit "valid" to be always true as below

```
size = lots_of_pkts.size();
```

```
for (i = 0; i < size; i++) begin
  data = lots_of_pkts[i].randomize() with { valid ==1 };
  valid = lots_of_pkts[i].valid;
  if (valid == TRUE) begin
    inject(data);
  end
end
```

269. Which of following coding styles are better and why?

1)
```
for (i=0; i < length*count; i++) begin
  a[i] = b[i];
end
```

2)
```
l_end = length * count;
for (i=0; i < l_end; i++) begin
  a[i] = b[i]
end
```

2) is better because in 1), there is a multiplication operation to done in each iteration to check for the loop limits while in 2) the multiplication is done only once and stored in a variable.

270. What is wrong in this code?
```
class ABC;
   local int var;
endclass
class DEF extends ABC;
  function new();
    var = 10;
  endfunction
endclass
```

var is local to class ABC and is not available in derived class. So it cannot be used in the derived class.

271. What is a virtual interface and where is it used?

A **virtual interface** is a variable that points to an actual interface. It is used in classes to provide a connection point to access the signals in an interface through the virtual interface pointer. The following example shows an actual interface bus_if that groups a set of bus signals. A *BusTransactor* class then defines a virtual interface of this type that is used to access all signals from this bus_if for driving a request or waiting for a grant signal. The top level test module which instantiates the physical interface will pass the handle of same to the BusTransactor class through constructor which gets assigned to the virtual interface pointer.

```
interface bus_if; // A bus interface
  logic req, grant;
  logic [7:0] addr, data;
endinterface

class BusTransactor; // Bus transactor class
  virtual bus_if bus; // virtual interface of type bus_if

  function new( virtual bus_if b_if );
    bus = b_if; // initialize the virtual interface
  endfunction

  task request(); // request the bus
    bus.req <= 1'b1;
  endtask

  task wait_for_bus(); // wait for the bus to be granted
    @(posedge bus.grant);
  endtask
endclass

module top;
  bus_if bif(); // instantiate interfaces, connect signals etc
  initial begin
    BusTransactor xactor;
    xactor = new(bif);   //pass interface to constructor
  end
endmodule
```

272. What is the concept of factory and factory pattern?

In object oriented programming, a factory is a method or a function that is used to create different objects of a prototype or a class. The different classes are registered with the factory and the factory method can create objects of any of the registered class types by

calling the corresponding constructor. This method of creating objects through factory instead of calling the constructor method directly is called factory pattern.
Using factory based object creation instead of calling constructors directly allows one to use polymorphism for object creation. This concept is implemented in UVM (Universal Verification Methodology) base class library and is used for creating and overriding base class objects with derived class objects.

273. What is the concept of callback?

A "callback" is any function that is called by another function which takes the first function as an argument. Most of the times, a callback function is called when some "event" happens. In a Verification testbench, this feature is useful for several applications:
1) Calling back a function to inject error on transactions sent from a driver
2) When a simulation phase is ready to end, calling a function to drain all pending transactions in all sequence/driver.
3) Calling a coverage sample function on a specific event.

Most of the times, callback functions are implemented by registering them with a component/object that calls back on some defined conditions.

An example call back function in UVM is phase_ready_to_end() which is implemented in the base class and is registered with the UVM_component class. The function gets called when the current simulation phase is ready to end always. Hence, a user can implement any functionality that needs to be executed at end of a simulation phase by overriding this function definition

274. What is a DPI call?

DPI stands for Direct Programming Interface and it is an interface between SystemVerilog and a foreign programming language like C/C++. DPI allows direct inter-language function calls between the languages on either side of the interface. Functions implemented in C language can be called in SystemVerilog (imported) and functions implemented in SystemVerilog can be called in C language (exported) using the DPI layer. DPI supports both functions (executing in zero time) and tasks (execution consuming time) across the language boundary. SystemVerilog data types are the only data types that can cross the boundary between SystemVerilog and a foreign language in either direction.

275. What is the difference between "DPI import" and "DPI export"?

A **DPI imported function** is a function that is implemented in the C language and called in the SystemVerilog code.
A **DPI exported function** is a function that is implemented in the SystemVerilog language and exported to C language such that it can be called from C language.
Both functions and tasks can be either imported or exported.

276. What are system tasks and functions? Give some example of system tasks and functions with their purpose.

SystemVerilog language supports a number of built-in system tasks and functions for different utilities and are generally called with a "$" prefix to the task/function name. In addition, language also supports addition of user defined system tasks and functions. Following are some examples of system tasks and functions (categorized based on functionality). For a complete list, one should refer to LRM.
1) Simulation control tasks - `$finish, $stop, $exit`
2) Conversion functions - `$bitstoreal, $itor, $cast`
3) Bit vector system functions - `$countones, $onehot, $isunknown`
4) Severity tasks - `$error, $fatal, $warning`
5) Sampled value system functions - `$rose, $fell, $changed`
6) Assertion control tasks - `$asserton, $assertoff`

Chapter 5: Fundamentals of Verification

In Digital VLSI Verification Interview, most of the times a candidate is given a simple digital design and corresponding design specification (Design Example: a full adder, or a simple ALU, or a simple cache, or a Multi-master bus, etc.). Candidate is then asked to define a verification strategy and explain the steps to verify that design. In this way, an interviewer can test a candidate for his/her general awareness and experience on verification as well as judge how well a candidate can think, analyze and solve given a problem statement. Following section contains questions that are designed to help you crack this segment of an interview. Additionally, this section also contains some commonly asked questions related to fundamentals of Verification. A Recent College Graduate may not be asked a lot of questions from this section, whereas this section may constitute a considerable portion of interview for a Senior candidate. In an interview, difficulty of this section varies with the experience of the candidate.

277. What is the difference between directed testing and constrained random verification? What are the advantages and disadvantage of both?

Directed testing is an approach where a directed test is written for verifying each of the features in the design. On the other hand, constrained random testing is an approach in which a stimulus is generated automatically using constrained random generators that generates stimulus as per design specification. Following table provides a comparison in terms of advantages and disadvantages of both. A recommended approach is to use a mix of both - constrained random to cover most of verification space and then directed tests to cover hard to reach corner cases.

Directed Testing	Constrained Random Testing
One or more directed tests are written to verify each feature of the design.	A stimulus generator is implemented that models constraints on stimulus and design specification to generate tests automatically
Provides good visibility and predictability to Verification progress as each test correlates to a design feature and hence can be tracked easily	Since tests are generated automatically, need extra effort to develop functional coverage monitors and collect coverage to ensure features are verified
Directed tests are easier to develop once the design features are understood	Developing constrained random verification testbenches is more complex and needs more experience. It can also take more time for development of verification environments.
For complex designs, writing and maintaining large directed test suites is very painful and time consuming	The constrained random generator is relatively easier to maintain once developed compared to large test suites.
Directed test writing is limited to scenarios that are identified by understanding the design specification.	Constrained random generator can cover more scenarios and features in combination along with random configuration to stress design better and cover some scenarios that might be missed by manual identification.

278. Explain what are self-checking tests?

A self-checking test is one that can check the result of the test by some means at the end of test. The results can be predicted in the test either by computing results of some memory operation or by gathering results from the DUT, like status registers or any other information.

279. What is Coverage driven verification?

In **Coverage driven Verification methodology**, a verification plan is implemented by mapping each of the feature or scenario into a coverage monitor that can be used to collect coverage information during simulation.
1) The coverage monitor can be a combination of sample based covergroups and property based coverage.
2) In coverage based verification, tests are normally generated using a constrained random stimulus generator, test correctness is ensured by functional checkers, and coverage is collected for all the coverage monitors implemented.

3) Usually multiple tests or multiple seeds of the random generator are regressed on the design and the individual coverage collected from each of the test is merged to get a cumulative coverage. Sometimes, a corner case in the design may not be covered easily using constrained random stimulus and might be better done using a directed test.
4) Coverage information also provides a feedback to the quality of tests and constraints in the generator and helps in fine tuning constraints for efficient random stimulus generation.
5) Since, in this approach the coverage definition is the key step on which the verification execution is tracked for progress and completion, it is important to make sure that the coverage definition and implementation is reviewed for completeness and correctness against verification plan and design specification.

280. What is the concept of test grading in functional verification?

Functional verification for a design is done in terms of creating directed tests as well as constrained random stimulus generator with different controls on stimulus. Through a design verification project, a set of tests gets developed and this test suite is used for verifying design correctness, finding bugs in design and for collecting coverage etc.

Test grading is a process in which individual tests are graded for quality in terms of different criteria like functional coverage hit, bugs found, simulation run time, ease of maintenance, etc.

This process helps in identifying efficient tests out of a test suite and thus developing the most efficient test suite for design verification.

281. What is Assertion based Verification (ABV) methodology?

Assertion-based verification (ABV) is a methodology in which assertions are used to capture specific design intent. These assertions are then used in simulation, formal verification, and/or emulation to verify if the design implementation is correct. ABV methodology can supplement other functional verification methodologies for efficient verification by making use of the benefits of assertions. Some of the benefits of assertions are following:
1) Assertions detect design errors at their source and thus help in increasing observability and decreasing debug time.
2) Same assertions can be used in both simulation and formal analysis and even in emulation.
3) A lot of assertions for general designs are available in assertion libraries and can be easily ported to any verification environment.
4) SystemVerilog Assertions written as properties can also be used for coverage (using cover properties), and hence help in coverage based verification methodology.

282. A 2x2 port packet switch has following specifications:

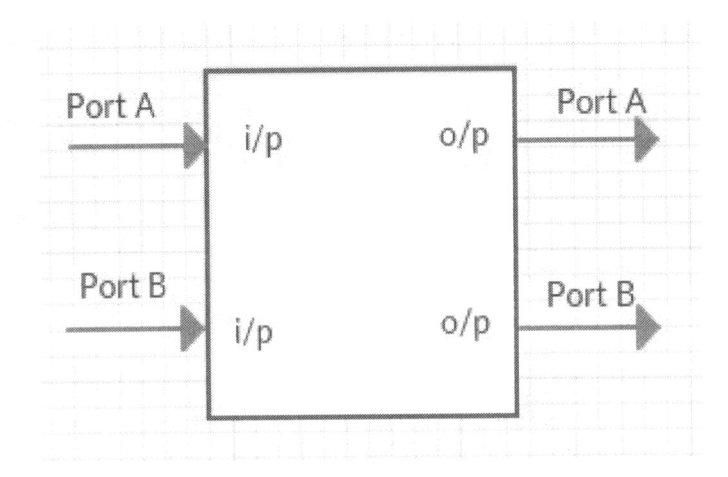

Specification: There are two input and output ports A and B as shown above. Each port can receive packets of variable size between 64 and 1518 bytes. Each packet will have a 4 Byte Source Address and 4 Byte destination address along with data and a 4 Byte CRC computed across packet as shown below. The packet will be switched to one of the output port based on Destination Address.

| CRC | Data | DA | SA |

How will you verify the design? How will you generate stimulus and checkers? What will be some of your corner cases for verification?

Answer:
For these kinds of questions where a design specification is given, first step is to understand the design specification and clarify any questions with the interviewer. The next step is to identify the scenarios to be verified and come up with a verification plan and strategy document. This should list down the features/scenarios to be verified, what methodologies can be used to verify (directed/constrained random, coverage, assertion, etc.), how to check for correctness etc. Further, details should be provided on how the stimulus can be generated and how checking can be done. Another aspect is to think through all design features and identify corner cases that need to be verified.
Now, let's try to list down how this simple router design can be verified
  1) Following are some of the scenarios that need to be verified:
     a) Test for proper switching of packets from Port A to both output ports based on destination address.
     b) Test for different packet sizes - minimum size, maximum size and random sizes in between will be good.
     c) Test for all possible values of Source and Destination Address.
     d) Test for different data patterns.
     e) Test for streaming packets (back to back with no delay, few cycles delay and large cycles delay), same size packets streamed or different size packets streamed.
     f) Test for correct CRC functionality by expecting the correct CRC.

g) Tests where some bits of SA/DA or data or even the CRC are corrupted.
h) What more can you think of?
2) Now, in order to verify above scenarios, we need to design a constrained random packet generator and we also need a scoreboard/checker that checks for packet correctness and correct switching behavior. If the tests are random, we will also need to write some coverage monitors that make sure all the important scenarios as mentioned above are getting covered.
3) If an interviewer wants to test you more, then he can also continue with questions asking you to actually write a SystemVerilog packet generator code or a checker or driver etc.

283. Given a RAM with a single port for read and write - what all conditions need to be verified?

A single port RAM has only a single port for read and write. So, it can only do a read or a write at any given point of time. Other design specifications that need to be considered for verification are the RAM size, and width of address and data bus.
Based on this, following are some of the scenarios that should to be verified:
1) Single Read and Write behaves correctly,
2) Back to Back reads or writes to same address and different addresses,
3) Writes followed by reads to same address back to back,
4) Reads followed by writes to same address back to back,
5) Verifying the boundary of RAM sizes - reads and writes,
6) Verifying the different patterns of writing into memory location like writing all zeros, all ones, alternating zero/one, walking one/zero patterns.

If you are further requested to define a verification environment, you can consider scenarios like above and define whether a directed or constrained random environment will work better and how the stimulus generator and checkers can be designed.

284. What is the difference between a single port RAM and dual port RAM?

A single port RAM has only a single port for read and write. So it can only do a read or a write at any given point of time. A dual port RAM has 2 ports for read/write and hence does allow read or write simultaneously.

285. A simple ALU with a block diagram as shown below supports two 4-bit operands, a 4-bit result bus, and carry overflow. The ALU supports up to 8 instructions using a 3-bit opcode or select lines (S2, S1, S0) with a decoding as below. Explain all the scenarios that needs to be verified to make sure the ALU works as per the specification table below:

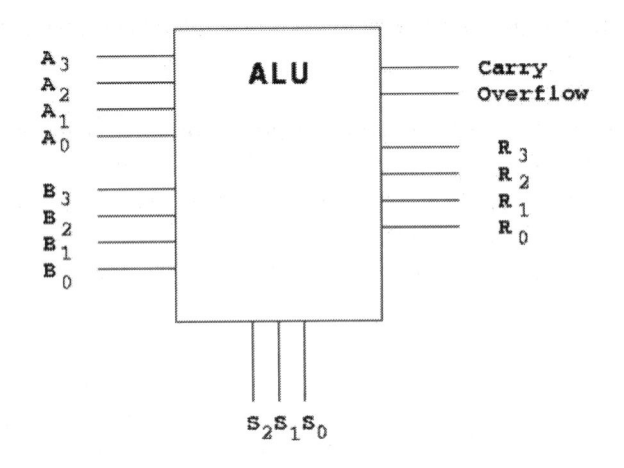

S2,S1,S0	Description
0	Add A, B (A+B)
1	Sub A, B (A-B)
10	Increment A (A=A+1)
11	Increment B (B=B+1)
100	A AND B (logical AND)
101	A OR B (logical OR)
110-111	Undefined and No operation

Answer:
Following are the scenarios that need to be verified for this given ALU design:
1) Verify that all individual operations work (Add, Sub, Increment, AND, and OR) by driving the two operands A and B, and driving the select lines for each of the operation.
2) Verify that if select lines are between 110-111, then no operation happens.
3) For each of above instructions, select minimum and maximum values of A and B and combinations. Given that A and B are 4 bit, maximum value possible is 4'b1111
4) Verify the overflow and underflow cases for ADD and SUB cases. If both A and B are 4'b1111, overflow happens for ADD, while if value of B is greater than A, underflow happens for SUB.
5) Verify wrap around cases for increment instructions. If A=4'b1111, increment should cause a value of 0.
6) Once individual scenarios are verified, create random sequence of opcodes to verify that effect of one operation does not affect the following ones. Check for sequences where same opcodes repeat more than once or different opcodes repeat in different patterns.
7) To create stimulus, you can design a random opcode and operand generator and a simple driver. To check for results, a simple model or ALU can be written and the results can be compared against same.

286. What is the difference between an event driven and a cycle based simulator?

Event Driven Simulators evaluate a design on every event, by taking each event and propagating the changes through design until a steady state condition is reached. An event is defined as a change in any of the input stimuli for a design element. A design element may be evaluated several times in a single cycle because of different arrival times of the inputs and the feedback of signals from downstream design elements.
For example: Consider a case of a logic path between two flip-flops operating on a clock. The combinational logic path can have several gates and feedback path. On a clock change, when the output of first flip-flop changes, it is applied on the input of the logic path and further any change at the input of the different stages in combinational logic, will trigger that specific design to be evaluated. This might take several evaluations before the value stabilizes and no longer change in that clock cycle. Most of the industry wide used simulators are event driven like: Questa from Mentor, VCS from Synopsys or Incisive Simulator from Cadence. This is because event driven simulators provides accurate simulation environment.
Cycle-Based Simulators have no notion of time within a clock cycle. They evaluate the logic between state elements and/or ports in a single shot. This helps in significant increase in simulation speeds as each logic element is evaluated only once per cycle. The disadvantage is that it cannot really detect any glitches in signals, and it works really well only on logic designs that are fully synchronous. Since timing of design is not taken into account during simulation, separate effort needs to be done on timing verification using any of the static timing analysis tools. Cycle based simulators are not very popular for general designs but are custom made and used at some of the companies that develops large designs like microprocessors.

287. What is a transaction? What are the benefits of Transaction based verification?

A transaction is a higher level abstraction of a group of low level information, like a group of signals. While designs operate at signal level information, testbenches need to have drivers and monitors interfacing at signal level with the design, while all other aspects of testbenches can be abstracted to be at a transaction level. Transaction based Verification is an approach in which a testbench is architected in a layered fashion where only lower layered components operate at signal level and all other components operate and communicate based on transactions as shown below.

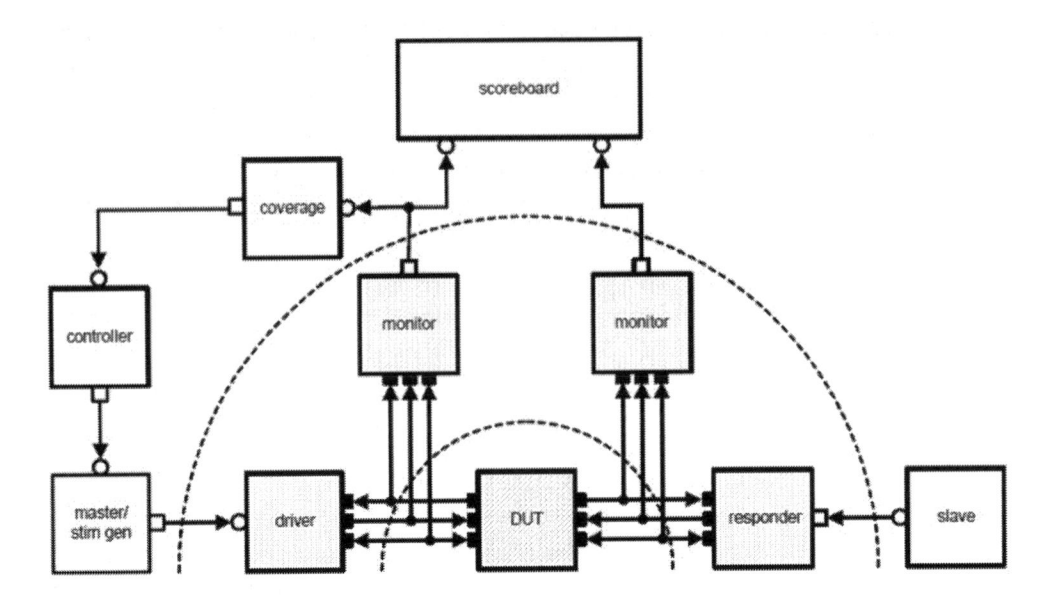

1) The main advantage of transaction based verification is in terms of re-using components developed with transactional interface in different verification environments within a project or across different projects. For example: with reference to above diagram, only the driver, monitor and responder needs to have a signal level interface. Once these components group signal level information to a transaction, other components like stimulus generators, slave models and scoreboards can all operate on transactions.
2) Since transactional components need to be evaluated by a simulator on a transactional boundary and not on every signal changes, simulations can be little faster.
3) If a design changes in terms of interface timing, then only the driver and monitor component need a change while other components will be unaffected.

288. What all simulation/debug tools have you worked on or are you familiar with?

This is a general question to test your awareness on different tools. Based upon your answer and experience with different tools, you could also be asked about your views in terms of easiness/limitations that you might have come across while using these tools. There is no fixed answer to this, but commonly used simulators are Questa from Mentor Graphics, VCS from Synopsys, and Incisive simulator from Cadence. Verdi from Synopsys is also a commonly used tool for debugging along-with DVE. Formal tools include Jasper from Cadence, and QuestaFormal from Mentor graphics.

289. When do we need reference model for verifying RTL designs? What are the advantages of using reference-models?

A reference model is usually a non-synthesizable model of the design specification that is usually written in a high level programming language like C/SystemVerilog. The reference model is sometimes implemented either to match design specification at a cycle level accuracy, or at a higher level boundary. For Example: a reference model of a CPU/microprocessor should be accurately modelling the state at an instruction boundary, while a reference model for an AMBA bus protocol should be cycle accurate as per the protocol.

Reference models are normally used in checkers/scoreboards to generate an expected response for a given stimulus pattern so that it can be compared against actual result or the output obtained from the design.

290. What is a Bus Functional Model?

Traditionally, Bus Functional Model (BFM) is a non-synthesizable model written in a high level programming language like C/SystemVerilog that models the functionality of a bus interface and can be connected to a Design interface for simulating the design. On one side of BFM, will be an interface that implements the bus protocol at signal level, and the other side will have an interface to support sending or receiving transactions.

Overtime this definition has evolved and in methodologies like UVM, there is no real component like a BFM, but the functionality is implemented by a collection of components like a driver, a monitor and a receiver.

291. How would you track the progress of the verification project? What metrics would you use?

A number of metrics are used to track the progress of verification against a plan. A verification plan captures the scenarios/features to be verified in terms of directed tests or in terms of functional coverage monitor for detailed scenarios and corner cases. The plan also captures details on verification environment development which includes stimulus generation and checking methodologies.

Progress can be tracked in early stage of project by tracking completeness of environment development (stimulus generator, checker, monitor etc.), test development and functional coverage monitor development. Once most of the tests and a constrained random generator is developed, then tests are normally run as regressions on a farm of servers, and then progress is monitored in terms of regression pass rates, bug rate and the functional coverage numbers.

292. How do you measure completeness of Verification OR when/how can you say that verification is complete?

Functional Verification can be called complete when the implemented behavior of a design matches with the design specification without any errors. To achieve this, we need to apply stimulus to the design to cover every possible input scenario and verify that the design meets specification without any errors. However, with ever increasing complexity of the designs, it is practically not possible to define all the possible input stimulus scenarios. In addition, resource and time limitations also make this ideal definition of completeness impractical.

Hence, in most of the projects, verification completeness is about the confidence achieved through a set of metrics and processes that minimizes the risk of having a design defect. Following are some of the metrics and processes that are followed to achieve high confidence with respect to verification completeness:

1) Reviewing Verification plan and design specification to make sure all details are understood and captured.
2) Ensuring proper completeness in terms of environment development, test development, and functional coverage monitor development against the reviewed plan.
3) Review of testbench stimulus generator and constraints, checkers, assertions and coverage monitor implementation.
4) Ensuring all tests are enabled in regression mode with consistently no failures across weeks, all coverage metrics met and understood.
5) Ensuring that bug rates and unresolved bugs are zero or well understood to have no impact on design.
6) Waveform Review of important scenarios.
7) Ensuring formal verification is done (wherever possible).
8) Comparing rate of incoming bugs and bug trend with that of past successful projects of similar complexity.

293. What is GLS and why is it important?

GLS is an acronym for "Gate Level Simulation". Gate Level Simulations are run after RTL code is synthesized into Gate Level Netlist. GLS forms an important part of Verification lifecycle. It is required in addition to static verification tools like STA (Static Timing Analysis) and LEC (Logical Equivalence Checking) as STA and LEC don't cover/report all issues. Majorly, GLS is used to:

1. Verify DFT scan chains.
2. Verify critical timing paths in asynchronous designs (this is not done by STA)
3. Verify Reset and Power Up flows.
4. Analyze X-Sate Optimism in RTL.
5. Collect Switching Activity for Power Estimation.

294. What are Power and Performance Trade-offs?

Power and performance are two important design points for a successful product. While most designs would ideally like to have highest possible performance with lowest possible power consumption, it is not practically possible always.

Dynamic power consumption is directly proportional to CV^2f, where f is the frequency, V is voltage, and C is capacitance. Hence, In general:

1) Decreasing Voltage will reduce power consumption but lowers performance (as delay increases)
2) Reducing Frequency will reduce power consumption but lowers performance (clock is slower)

Hence, for an optimal performance and power target, design needs to make a choice of right Voltage and frequency values.

Note: More questions on power and clocking are present in "Power and Clocking" Section (6.3) in the next chapter (Verification Methodologies).

Chapter 6: Verification Methodologies

It is a known fact that Functional Verification consumes significant amount of time and effort in an overall product lifecycle. With availability of several tools and techniques, defining and deciding on verification methodologies that could enable fast and efficient execution towards a bug free design is becoming an important step of Verification planning phase. Verification methodologies include Dynamic Simulation vs Formal Verification, Assertion based Verification, Coverage methodology, Power Aware Simulations, Performance Verification and also UVM (Universal Verification Methodology) for constrained random testbenches.

This section is organized into several subsections (covering each of these methodologies), and various relevant concepts are explained through detailed answers for commonly asked Interview questions.

6.1 UVM (Universal Verification Methodology)

UVM is a standard verification methodology which is getting more and more popularity and wider adoption in verification industry. The methodology was created by Accellera and is currently in the IEEE working group 1800.2 for standardization. UVM consists of a defined methodology in terms of architecting testbenches and test cases and also comes with a library of classes that helps in building efficient constrained random testbenches easily.

This section has questions that test your general understanding of UVM methodology and details on usage of UVM in building constrained random testbenches.

295. What are some of the benefits of UVM methodology?

UVM is a standard verification methodology which is getting standardized as IEEE1800.2 standard. UVM consists of a defined methodology in terms of architecting testbenches and test cases, and also comes with a library of classes that helps in building efficient constrained random testbenches easily.
Some of the advantages and focus of the methodology include following:
1) Modularity and Reusability - The methodology is designed as modular components (Driver, Sequencer, Agents, Env, etc.) and this enables reusing components across unit level to multi-unit or chip level verification as well as across projects.
2) Separating Tests from Testbenches - Tests in terms of stimulus/sequencers are kept separate from the actual testbench hierarchy and hence stimulus can be reused across different units or across projects.
3) Simulator Independent - The base class library and the methodology is supported by all simulators and hence there is no dependence on any specific simulator.
4) Sequence methodology gives good control on stimulus generation. There are several ways in which sequences can be developed: randomization, layered sequences, virtual sequences, etc. This provides a good control and rich stimulus generation capability.

5) Config mechanisms simplify configuration of objects with deep hierarchy. The configuration mechanism helps in easily configuring different testbench components based upon verification environment using it, and without worrying about how deep any component is in the testbench hierarchy.
6) Factory mechanisms simplify modification of components easily. Creating each components using factory enables them to be overridden in different tests or environments without changing underlying code base.

296. What are some of the drawbacks of UVM methodology?

With increasing adoption of UVM methodology in the verification industry, it should be clear that the advantages of UVM overweight any drawbacks.
1) For anyone new to the methodology, the learning curve to understand all details and the library is very steep.
2) The methodology is still developing and has a lot of overhead that can sometimes cause simulation to appear slow or probably can have some bugs

297. What is the concept of Transaction Level Modelling?

Transaction level Modelling (TLM) is an approach to model any system or design at a higher level of abstraction. In TLM, communication between different modules is modelled using Transactions thus abstracting away all low level implementation details. This is one of the key concepts used in verification methodologies to increase productivity in terms of modularity and reuse. Even though the actual interface to the DUT is represented by signal-level activity, most of the verification tasks such as generating stimulus, functional checking, collecting coverage data, etc. are better done at transaction level by keeping them independent of actual signal level details. This helps those components to be reused and better maintained within and across projects.

298. What are TLM ports and exports?

In Transaction Level Modelling, different components or modules communicate using transaction objects. A TLM port defines a set of methods (API) used for a particular connection while the actual implementation of these methods are called TLM exports. A connection between the TLM port and the export establishes a mechanism of communication between two components.

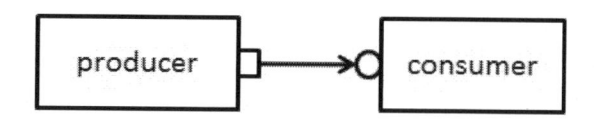

Here is a simple example of how a producer can communicate to a consumer using a simple TLM port. The producer can create a transaction and "put" to the TLM port, while the implementation of "put" method which is also called TLM export would be in the consumer that reads the transaction created by producer, thus establishing a channel of communication.

299. What are TLM FIFOs?

A TLM FIFO is used for Transactional communication if both the producing component and the consuming component need to operate independently. In this case (as shown below), the producing component generates transactions and "puts" into FIFO, while the consuming component gets one transaction at a time from the FIFO and processes it.

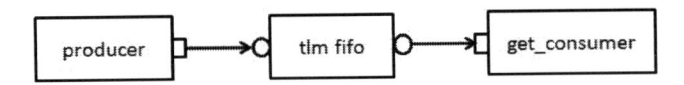

```
producer  →  tlm fifo  →  get_consumer
```

300. What is the difference between a get() and peek() operation on a TLM fifo?

The `get()` operation will return a transaction (if available) from the TLM FIFO and also removes the item from the FIFO. If no items are available in the FIFO, it will block and wait until the FIFO has at least one entry.
The `peek()` operation will return a transaction (if available) from the TLM FIFO without actually removing the item from the FIFO. It is also a blocking call which waits if FIFO has no available entry.

301. What is the difference between a get() and try_get() operation on a TLM fifo?

`get()` is a blocking call to get a transaction from TLM FIFO. Since it is blocking, the task `get()` will wait if no items are available in the FIFO stalling execution. On the other hand, `try_get()` is a nonblocking call which will return immediately even if no items are available in the FIFO. The return value of `try_get()` indicates if a valid item is returned or not.
Following are two equivalent implementations using `get()` and `try_get()`
 1) Using the blocking method - get()

```
class consumer extends uvm_component;
  uvm_get_port #(simple_trans) get_port;
  task run;
    for(int i=0; i<10; i++) begin
      t = get(); //blocks until a transaction is returned
      //Do something with it.
    end
```

```
        endtask
      endclass
```

2) Equivalent implementation using nonblocking method - try_get()

```
    class consumer extends uvm_component;
      uvm_get_port #(simple_trans) get_port;
      task run;
        for(int i=0; i<10; i++) begin
          //Try get is nonblocking. So keep attempting
          //on every cycle until you get something
          //when it returns true
          while(!get_port.try_get(t)) begin
            wait_cycle(1); //Task that waits one clock cycle
          end
          //Do something with it
        end
      endtask
    endclass
```

302. What is the difference between analysis ports and TLM ports? And what is the difference between analysis FIFOs and TLM FIFOs? Where are the analysis ports/FIFOs used?

The TLM ports/FIFOs are used for transaction level communication between two components that have a communication channel established using put/get methods. Analysis ports/FIFOs are another transactional communication channel which are meant for a component to distribute (or broadcast) transaction to more than one component. TLM ports/FIFOs are used for connection between driver and sequencer while analysis ports/FIFOs are used by monitor to broadcast transactions which can be received by scoreboard or coverage collecting components.

303. What is the difference between a sequence and sequence item?

A sequence item is an object that models the information being transmitted between two components (sometimes it can also be called a transaction). For Example: consider memory access from a CPU to the main memory where CPU can do a memory read or a memory write, and each of the transaction will have some information like the address, data and read/write type.

A sequence can be thought of a defined pattern of sequence items that can be send to the driver for injecting into the design. The pattern of sequence items is defined by how the body() method is implemented in sequence. For Example: Extending above example, we can define a sequence of 10 transactions of reads to incremental memory addresses. In this case, the body() method will be implemented to generate sequence items 10 times, and send them to driver while say incrementing or randomizing address before next item.

304. What is the difference between a uvm_transaction and a uvm_sequence_item?

uvm_transaction is the base class for modelling any transaction which is derived from uvm_object.

A sequence item is nothing but a transaction that groups some information together and also adds some other information like: sequence id (id of sequence which generates this item), and transaction id (the id for this item), etc. It is recommended to use uvm_sequence_item for implementing sequence based stimulus.

305. What is the difference between copy(), clone(), and create() method in a component class?

1) The create() method is used to construct an object.
2) The copy() method is used to copy an object to another object.
3) The clone() method is a one-step command to create and copy an existing object to a new object handle. It will first create an object by calling the create() method and then calls the copy() method to copy existing object to the new handle.

306. Explain the concept of Agent in UVM methodology.

UVM agent is a component that collects together a group of other uvm_components focused around a specific pin-level interface for a DUT. Most of the DUTs have multiple logical interfaces and an Agent is used to group all: driver, sequencer, monitor, and other components, operating at that specific interface. Organizing components using this hierarchy helps in reusing an "Agent" across different verification environments and projects that have same interface.

Following diagram shows usually how a group of components are organized as agent.

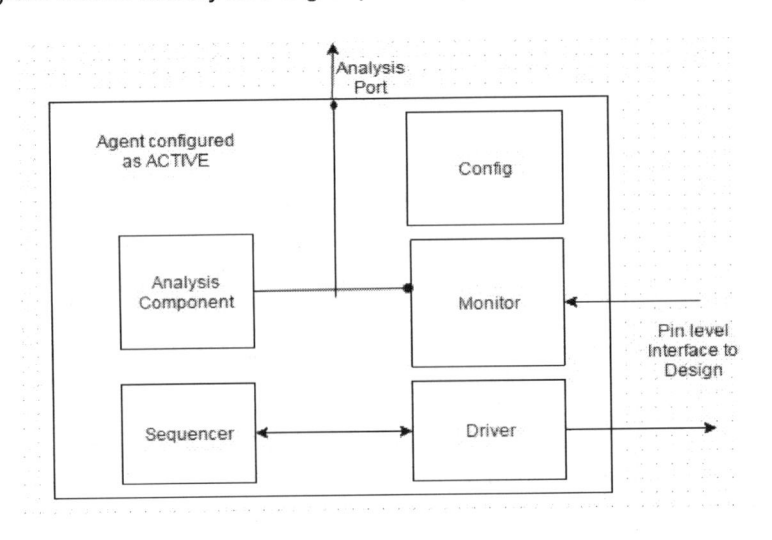

307. What all different components can a UVM agent have?

As explained in previous question, an agent is a collection of components that are grouped based on a logical interface to the DUT. An agent normally has a driver and a sequencer to drive stimulus to the DUT on the interface on which it operates. It also has a monitor and an analysis component (like a scoreboard or a coverage collector) to analyze activity on that interface. In addition, it can also have a configuration object that configures the agent and its components.

308. What is the difference between get_name() and get_full_name() methods in a uvm_object class?

The get_name() function returns the name of an object, as provided by the name argument in the new constructor or set_name() method.
The get_full_name() returns the full hierarchical name of an object. For uvm_components, this is useful when used in print statements as it shows the full hierarchy of a component. For sequence or config objects that don't have a hierarchy, this prints the same value as a get_name()

309. How is ACTIVE agent different from PASSIVE agent?

An **ACTIVE** agent is an agent that can generate activity at the pin level interface on which it operates. This means, the components like driver and sequencer would be connected and there would be a sequence running on it to generate activity.
A **PASSIVE** agent is an agent that doesn't generate any activity but can only monitor activity happening on the interface. This means, in a passive agent the driver and sequencer will not be created.
An Agent is normally configured ACTIVE in a block level verification environment where stimulus is required to be generated. Same agent can be configured PASSIVE as we move from block level to chip level verification environment in which no stimulus generation is needed, but we still can use same for monitoring activity in terms of debug or coverage.

310. How is an Agent configured as ACTIVE or PASSIVE?

UVM agents have a variable of type UVM_ACTIVE_PASSIVE_e which defines whether the agent is active (UVM_ACTIVE) with the sequencer and the driver constructed, or passive (UVM_PASSIVE) with neither the driver nor the sequencer constructed. This parameter is called active and by default it is set to UVM_ACTIVE.
This can be changed using set_config_int() while the agent is created in the environment class. The build phase of the agent should then have the code as below to selectively construct driver and sequencer.
```
function void build_phase(uvm_phase phase);
```

```
      if(m_cfg.active == UVM_ACTIVE) begin
        //create driver, sequencer
      end
   endfunction
```

311. What is a sequencer and a driver, and why are they needed?

A **Driver** is a component that converts a transaction or sequence item into a set of pin level toggling based on the signal interface protocol.
A **Sequencer** is a component that routes sequence items from a sequence to a driver and routes responses back from driver to sequence. The sequencer also takes care of arbitration between multiple sequences (if present) trying to access driver to stimulate the design interface.
These components are needed as in a TLM methodology like UVM, stimulus generation is abstracted in terms of transactions and the sequencer and driver are the components that route them and translate them to actual toggling of pins.

312. What is the difference between a monitor and a scoreboard in UVM?

A monitor is a component that observes pin level activity and converts its observations into transactions or sequence_items. It also sends these transactions to analysis components through an analysis port.
A scoreboard is an analysis component that checks if the DUT is behaving correctly. UVM scoreboards use analysis transactions from the monitors implemented inside agents.

313. Which method activates UVM testbench and how is it called?

The `run_test()` method (a static method) activates the UVM testbench. It is normally called in an "initial begin ... end" block of a top level test module, and it takes an argument that defines the test class to be run. It then triggers construction of test class and the `build_phase()` will execute and further construct Env/Agent/Driver/Sequencer objects in the testbench hierarchy.

314. What steps are needed to run a sequence?

There are three steps needed to run a sequence as follows:
1) Creating a sequence. A sequence is created using the factory create method as shown below:
     ```
     my_sequence_c  seq;
     seq = my_sequence_c::type_id::create("my_seq")
     ```
2) Configuring or randomizing a sequence. A sequence might have several data members that might need configuration or randomization. Accordingly, either configure values or call

```
        seq.randomize()
```
3) Starting a sequence. A sequence is started using `sequence.start()` method. The start method takes an argument which is the pointer to the sequencer on which sequence has to be run. Once the sequence is started, the `body()` method in the sequence gets executed and it defines how the sequence operates. The `start()` method is blocking and returns only after the sequence completes execution.

315. Explain the protocol handshake between a sequencer and driver?

The UVM sequence-driver API majorly uses blocking methods on sequence and driver side as explained below for transferring a sequence item from sequence to driver and collecting response back from driver.

On the sequence side, there are two methods as follows:
1) `start_item(<item>)`: This requests the sequencer to have access to the driver for the sequence item and returns when the driver grants access to the sequencer.
2) `finish_item(<item>)`: This method results in the driver receiving the sequence item and is a blocking method which returns only after driver calls the `item_done()` method.

On the driver side,
1) `get_next_item(req)`: This is a blocking method in driver that blocks until a sequence item is received on the port connected to sequencer. This method returns the sequence item which can be translated to pin level protocol by the driver.
2) `item_done(req)`: The driver uses this nonblocking call to signal to the sequencer that it can unblock the sequences *finish_item()* method, either when the driver accepts the sequences request or it has executed it.

Following diagram illustrates this protocol handshake between sequencer and driver which is the most commonly used handshake to transfer requests and responses between sequence and driver.

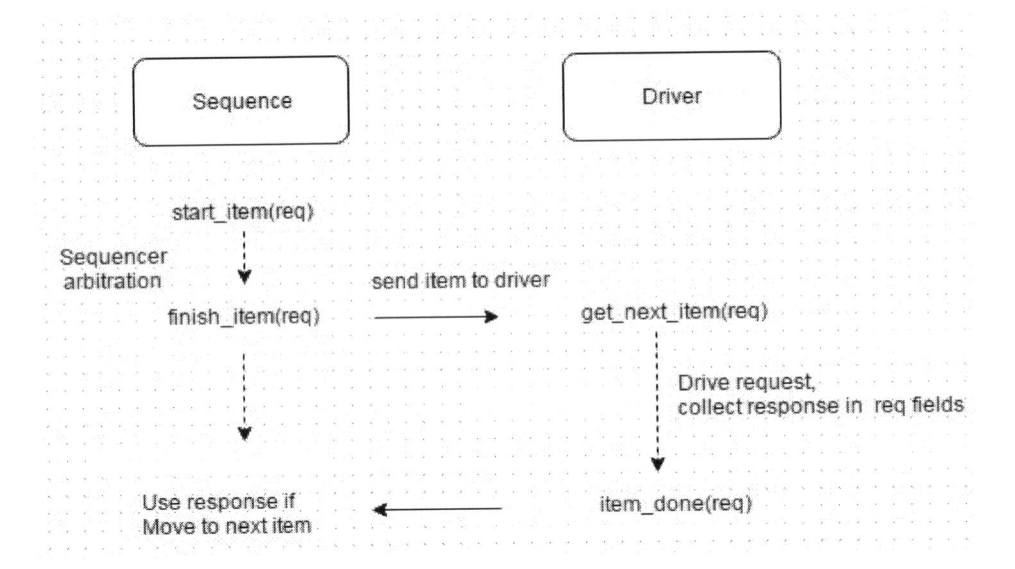

```
                Sequence                          Driver


            start_item(req)
   Sequencer        ┆
   arbitration      ▼         send item to driver
            finish_item(req)  ─────────────►    get_next_item(req)
                    ┆                                  ┆
                    ┆                                  ┆  Drive request,
                    ┆                                  ┆  collect response in req fields
                    ▼                                  ▼
            Use response if   ◄─────────────    item_done(req)
            Move to next item
```

Few other alternatives methods are: `get()` method in driver which is equivalent to calling `get_next_item()` along with `item_done()`. Sometimes, there would also be need for a separate response port if the response from driver to sequence contains more information than what could be encapsulated in the request class. In this case, sequencer will use a `get_response()` blocking method that gets unblocked when the driver sends a separate response on this port using put() method. This is illustrated in below diagram.

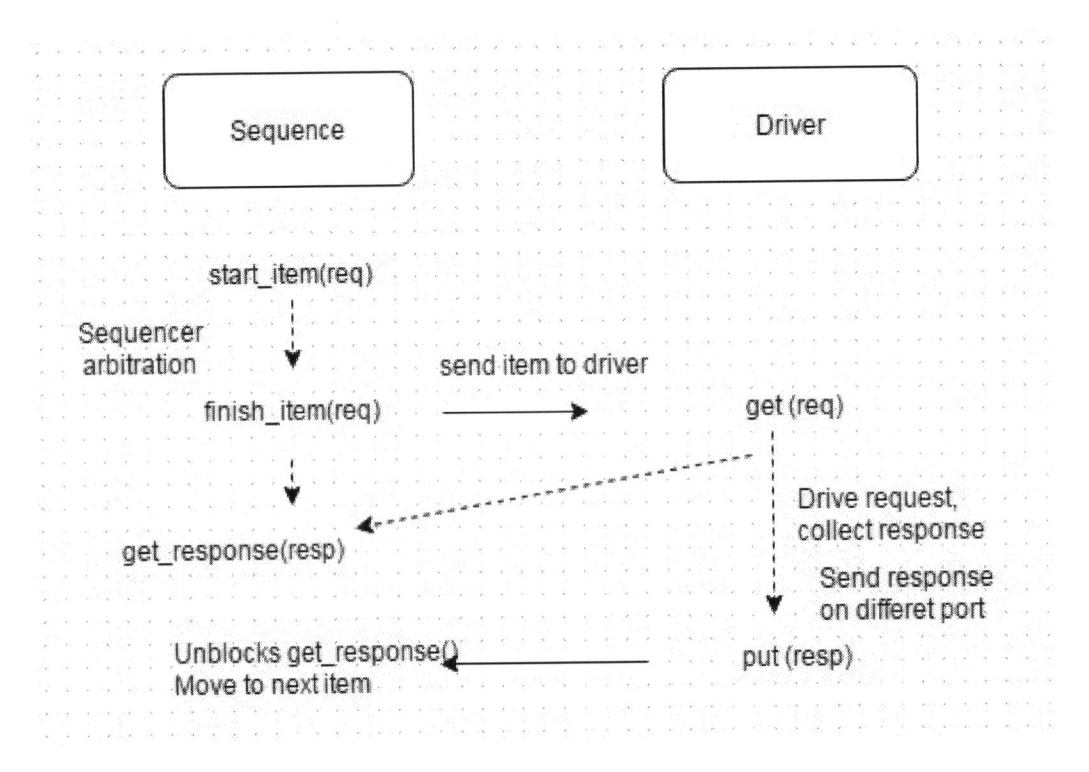

```
                Sequence                          Driver


            start_item(req)
   Sequencer        ┆
   arbitration      ▼         send item to driver
            finish_item(req)  ─────────────►    get (req)
                    ┆                                  ┆
                    ▼                                  ┆  Drive request,
            get_response(resp) ◄╌╌╌╌╌╌╌╌╌╌            ┆  collect response
                                                       ┆  Send response
                                                       ▼  on differet port
            Unblocks get_response() ◄─────────    put (resp)
            Move to next item
```

158

316. What are `pre_body()` and `post_body()` functions in a sequence? Do they always get called?

`pre_body()` is a method in a sequence class that gets called before the `body()` method of a sequence is called. `post_body()` method in sequence gets called after the `body()` method is called.

The `pre_body()` and `post_body()` methods are not always called. The uvm_sequence::start() has an optional argument which if set to 0, will result in these methods not being called. Following are the formal argument of start() method in a sequence.

```
virtual task start (
  uvm_sequencer_base sequencer, // Pointer to sequencer
  uvm_sequence_base parent_sequencer = null, // parent sequencer
  integer this_priority = 100, // Priority on the sequencer
  bit call_pre_post = 1); // pre_body and post_body called
```

317. Is the start() method on a sequence blocking or nonblocking?

The `start()` method is a blocking call. It blocks execution until the `body()` method of the sequence completes execution.

318. What are the different arbitration mechanisms available for a sequencer?

Multiple sequences can interact concurrently with a driver connected to a single interface. The sequencer supports an arbitration mechanism to ensure that at any point of time only one sequence has access to the driver. The choice of which sequence can send a sequence_item is dependent on a user selectable sequencer arbitration algorithm. There are five built-in sequencer arbitration mechanisms that are implemented in UVM. There is also an additional hook to implement a user defined algorithm.

The sequencer has a method called `set_arbitration()` that can be called to select which algorithm the sequencer should use for arbitration. The six algorithms that can be selected are following:

1) <u>SEQ_ARB_FIFO</u> (Default if none specified). If this arbitration mode is specified, then the sequencer picks sequence items in a FIFO order from all sequences running on the sequencer. For Example: if seq1, seq2 and seq3 are running on a sequencer, it will pick an item from seq1 first, followed by seq2, and then seq3 if available, and continue.
2) <u>SEQ_ARB_WEIGHTED:</u> If this arbitration mode is selected, sequence items from the highest priority sequence are always picked first until none available, then the

sequence items from next priority sequence, and so on. If two sequences have equal priority, then the items from them are picked in a random order.
3) <u>SEQ_ARB_RANDOM</u>: If this arbitration mode is selected, sequence items from different sequences are picked in a random order by ignoring all priorities.
4) <u>SEQ_ARB_STRICT_FIFO:</u> This is similar to SEQ_ARB_WEIGHTED except that if two sequences have same priority, then the items from those sequences are picked in a FIFO order rather than in a random order.
5) <u>SEQ_ARB_STRICT_RANDOM</u>: This is similar to SEQ_ARB_RANDOM except that the priorities are NOT ignored. The items are picked randomly from sequences with highest priority first followed by next and in that order.
6) <u>SEQ_ARB_USER:</u> This algorithm allows a user to define a custom algorithm for arbitration between sequences. This is done by extending the uvm_sequencer class and overriding the user_priority_arbitration() method.

319. How do we specify the priority of a sequence when it is started on a sequencer?

The priority is specified by passing an argument to the `start()` method of the sequence. The priority is decided based on relative values specified for difference sequences. For Example: If two sequences are started as follows, the third argument specifies the priority of the sequence.

```
seq_1.start(m_sequencer, this, 500); //Highest priority
seq_2.start(m_sequencer, this, 300); //Next Highest priority
seq_3.start(m_sequencer, this, 100); //Lowest priority among three
sequences
```

320. How can a sequence get exclusive access to a sequencer?

When multiple sequences are run on a sequencer, the sequencer arbitrates and grants access to each sequence on a sequence item boundary. Sometimes a sequence might want exclusive access to sequencer until all the sequence items part of it are driven to driver (for example: if you want to stimulate a deterministic pattern without any interruption). There are two mechanisms that allow a sequence to get exclusive access to sequencer.

1) **Using lock() and unlock()**: A sequence can call the lock method of the sequencer on which it runs. The calling sequence will be granted exclusive access to the driver when it gets the next slot via the sequencer arbitration mechanism. If there are other sequences marked as higher priority, this sequence needs to wait until it gets it slot. Once the lock is granted, no other sequences will be able to access the driver until the sequence issues an unlock() call on the sequencer which will then release the lock. The lock method is blocking and does not return until lock has been granted.

2) **Using grab() and ungrab()**: The grab method is similar to the lock method and can be called by the sequence when it needs exclusive access. The difference between

grab and lock is that when grab() is called, it takes immediate effect and the sequence will grab the next sequencer arbitration slot, overriding any sequence priorities in place. The only thing that can stop a sequence from grabbing a sequencer is an already existing lock() or grab() condition on the sequencer.

321. What is the difference between a grab() and a lock() on sequencer?

Both grab() and lock() methods of a sequencer are used by a sequence running on that sequencer to get exclusive access to the sequencer until the corresponding unlock() or ungrab() is called. The difference between grab and lock is that when a grab() on sequencer is called, it takes immediate effect and the sequence will grab the next sequencer arbitration slot overriding any sequence priorities in place. However, a call to lock() sequencer will need to wait until the calling sequence gets its next available slot based on priorities and arbitration mechanisms that are set.

In terms of usage, one example usage will be to use a lock to model a prioritised interrupt and a grab to model a non-maskable interrupt, but there are several other modelling scenarios where this will be useful as well.

322. What is the difference between a pipelined and a non-pipelined sequence-driver model?

Based on how a design interface needs to be stimulated, there can be two modes implemented in an UVM driver class.

1) **Non-pipelined model:** If the driver models only one active transaction at a time, then it is called a non-pipelined model. In this case, sequence can send one transaction to the driver and driver might take several cycles (based on the interface protocol) to finish driving that transaction. Only after that the driver will accept a new transaction from sequencer

```
class nonpipe_driver extends uvm_driver #(req_c);
  task run_phase(uvm_phase phase);
    req_c req;
    forever begin
      get_next_item(req); // Item from sequence via sequencer
      // drive request to DUT which can take more clocks
      // Sequence is blocked to send new items til then
      item_done(); // ** Unblocks finish_item() in sequence
    end
  endtask: run_phase
endclass: nonpipe_driver
```

2) **Pipelined model**: If the driver models more than one active transaction at a time, then it is called a pipelined model. In this case sequence can keep sending new transactions to driver without waiting for driver to complete a previous transaction. In this case, on every transaction send from the sequence, driver will fork a separate process to drive the interface signals based on that transaction, but will not wait until

it is completed before accepting a new transaction. This modelling is useful if we want to drive back to back requests on an interface without waiting for responses from design.

```
class pipeline_driver extends uvm_driver #(req_c);
  task run_phase(uvm_phase phase);
    req_c req;
    forever begin
      get_next_item(req); // Item from sequence via sequencer
      fork
      begin
        //drive request to DUT which can take more clocks
        //separate thread that doesn't block sequence
        //driver can accept more items without waiting
      end
      join_none
      item_done(); // ** Unblocks finish_item() in sequence
    end
  endtask: run_phase
endclass: pipeline_driver
```

323. How do we make sure that if multiple sequences are running on a sequencer-driver, responses are send back from driver to the correct sequence?

If responses are returned from the driver for one of several sequences, the sequence id field in the sequence is used by the sequencer to route the response back to the right sequence. The response handling code in the driver should use the *set_id_info()* call to ensure that any response items have the same sequence id as their originating request.
Here is an example code in driver that gets a sequence item and sends a response back
(*Note that this is a reference pseudo code for illustration and some functions are assumed to be coded somewhere else*)

```
class my_driver extends uvm_driver;
  //function that gets item from sequence port and
  //drives response back
  function  drive_and_send_response();
    forever begin
      seq_item_port.get(req_item);
      //function that  takes req_item and drives pins
      drive_req(req_item);
      //create a new response item
      rsp_item = new();
      //some function that monitors response signals from dut
      rsp_item.data =  m_vif.get_data();
      //copy id from req back to response
```

```
            rsp.set_id_info(req_item);
          //write response on rsp port
            rsp_port.write(rsp_item);
        end
      endfunction
    endclass
```

324. What is m_sequencer handle?

When a sequence is started, it is always associated with a sequencer on which it is started. The m_sequencer handle contains the reference to the sequencer on which sequence is running. Using this handle, the sequence can access any information and other resource handles in the UVM component hierarchy.

325. What is a p_sequencer handle and how is it different in m_sequencer?

A UVM sequence is an object with limited life time unlike a sequencer or a driver or a monitor which are UVM components and are present throughout simulation time. So if you need to access any members or handles from the testbench hierarchy (component hierarchy), the sequence would need a handle to the sequencer on which it is running.

m_sequencer is a handle of type uvm_sequencer_base which is available by-default in a uvm_sequence. However, to access the real sequencer on which sequence is running, we need to typecast the m_sequencer to the real sequencers, which is generally called p_sequencer (*though you could really use any name and not just p_sequencer*).

Here is a simple example where a sequence wants to access a handle to a clock monitor component which is available as a handle in the sequencer.

```
    class test_sequence_c extends uvm_sequence;
      test_sequencer_c p_sequencer;
      clock_monitor_c my_clock_monitor;

      task pre_body();
        if(!$cast(p_sequencer, m_sequencer)) begin
          `uvm_fatal("Sequencer Type Mismatch:", " Wrong Sequencer");
        end
        my_clock_monitor = p_sequencer.clk_monitor;
      endtask
    endclass
```

```
class test_Sequencer_c extends uvm_sequencer;
    clock_monitor_c clk_monitor;
endclass
```

326. What is the difference between early randomization and late randomization while generating a sequence?

In **Early randomization**, a sequence object is first randomized using randomize() call and then the sequence calls `start_item()` to request access to sequencer, which is a blocking call and can take time based upon how busy the sequencer is. Following example shows an object (req) randomized first and then sequence waits for arbitration

```
task body();
    assert(req.randomize());
    start_item(req);   //Can consume time based on sequencer arbitration
    finish_item(req);
endtask
```

In **Late randomization**, a sequence first calls `start_item()`, waits until arbitration is granted from the sequencer, and then just before sending the transaction to sequencer/driver, randomize is called. This has the advantage that items are randomized just in time and can use any feedback from design or other components just before sending an item to driver. Following code shows late randomization of a request (req)

```
task body();
    start_item(req);   //Can consume time based on sequencer arbitration
    assert(req.randomize());
    finish_item(req);
endtask
```

327. What is a subsequence?

A subsequence is a sequence that is started from another sequence. From the `body()` of a sequence, if `start()` of another sequence is called, it is generally called a subsequence.

328. What is the difference between `get_next_item()` and `try_next_item()` methods in UVM driver class?

The `get_next_item()` is a blocking call (part of the driver-sequencer API) which blocks until a sequence item is available for driver to process, and returns a pointer to the sequence item.

The `try_next_item()` is a nonblocking version which returns a null pointer if no sequence item is available for driver to process.

329. What is the difference between `get_next_item()` and `get()` methods in UVM driver class?

The `get_next_item()` is a blocking call to get the sequence item from the sequencer FIFO for processing by driver. Once the sequence item is processed by driver, it needs to call `item_done()` to complete the handshake before a new item is requested using `get_next_item()`.

The `get()` is also a blocking call which gets the sequence item from sequencer FIFO for processing by driver. However, while using get(), there is no need to explicitly call `item_done()` as the get() method completes the handshake implicitly.

330. What is the difference between get() and peek() methods of UVM driver class?

The get() method part of driver class is a blocking call which gets the sequence item from sequencer FIFO for processing by driver. It unblocks once an item is available and completes handshake with sequencer.

The peek() method is similar to get() and blocks until a sequence item is available. However, it will not remove the request from the sequencer FIFO. So calling peek() multiple times will return same sequencer item in driver.

331. What is the difference in item_done() method of driver-sequencer API when called with and without arguments?

The `item_done()` method is a nonblocking method in driver class that is used to complete handshake with the sequencer after a `get_next_item()` or `try_next_item()` is successful.

If there is no need to send a response back, `item_done()` is called with no argument which will complete the handshake without placing anything in the sequencer response FIFO.

If there is a need to send a response back, `item_done()` is passed with a pointer to a response sequence_item as an argument. This response pointer will be placed in the sequencer response FIFO which can be processed by the sequence as a response to the request it drove.

332. Which of the following driver class methods are blocking calls and which are nonblocking?

1) get()
2) get_next_item()
3) item_done()
4) put()
5) try_next_item()
6) peek()

get(), get_next_item(), peek() are blocking calls.
try_next_item(), item_done(), and put() are nonblocking calls

333. Which of the following code is wrong inside a UVM driver class?

1)
```
function get_drive_req();
  forever begin
    req = get();
    req = get();
  end
endfunction
```

2)
```
function get_drive_req();
  forever begin
    req = get_next_item();
    req = get_next_item();
    item_done();
  end
endfunction
```

3)
```
function get_drive_req();
  forever begin
    req = peek();
    req = peek();
    item_done();
    req = get();
  end
endfunction
```

2) is wrong as you cannot call *get_next_item()* twice before calling *item_done()* as it will not complete handshake with the sequencer.

334. How can you stop all sequences running on a sequencer?

The sequencer has a method `stop_sequences()` that can be used to stop all sequences. However, this method does not check if the driver is currently processing any sequence_items. Because of this, if driver calls an item_done() or put(), there can be a Fatal Error as the sequence pointer might not be valid. So a user needs to take care of making sure that once stop_sequence() is called, the sequencer thread is disabled (if started in a fork).

335. Which method in the sequence gets called when user calls sequence.print() method?

convert2string() : It is recommended to implement this function which returns a string representation of the object (values of its data members). This is useful to get debug information printed to simulator transcript or log file.

336. Identify any potential issues in following code part of a UVM sequence

```
task body();
    seq_item_c req;
    start_item(req);
    #10 ns;
    assert(req.randomize());
    finish_item(req);
endtask
```

Adding a delay between start_item and finish_item should be avoided. The sequence wins arbitration and has access to sequencer/driver once start_item returns. Any delay from then till finish_item will hold up the sequencer/driver and will not be available for any other sequence. This will be more problematic if multiple sequences are run on an interface and more the delay, more idle on the design interface.

337. What is a virtual sequence and where do we use a virtual sequence? What are its benefits?

A virtual sequence is a sequence which controls stimulus generation across multiple sequencers. Since sequences, sequencers and drivers are focused on single interfaces, almost all testbenches require a virtual sequence to co-ordinate the stimulus and the interactions across different interfaces. Virtual sequences are also useful at a Subsystem or System level testbenches to have unit level sequences exercised in a co-ordinated fashion.

Following diagram shows this conceptually where a Virtual sequence has handles to three sequencers which connect to drivers for three separate interface to DUT. The virtual sequence can then generate sub sequences on each of the interfaces and run them on the corresponding sub-sequencer.

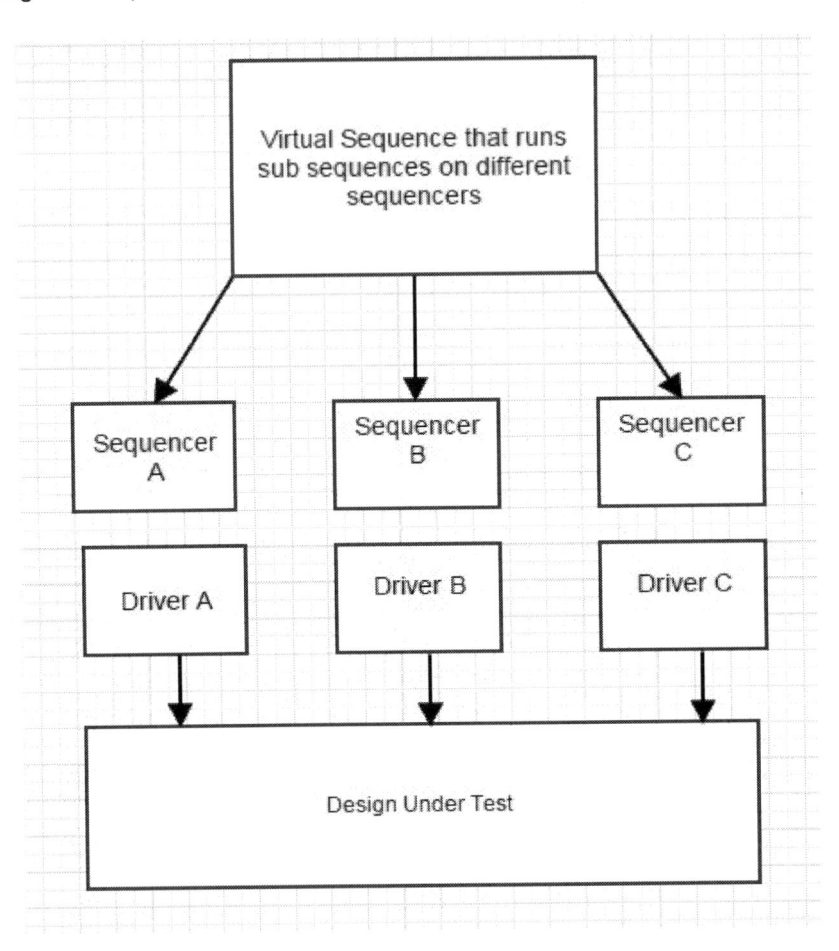

```
                    ┌─────────────────────────┐
                    │ Virtual Sequence that runs│
                    │ sub sequences on different│
                    │       sequencers         │
                    └─────────────────────────┘
                      │          │          │
                      ▼          ▼          ▼
                ┌─────────┐ ┌─────────┐ ┌─────────┐
                │Sequencer│ │Sequencer│ │Sequencer│
                │    A    │ │    B    │ │    C    │
                └─────────┘ └─────────┘ └─────────┘

                ┌─────────┐ ┌─────────┐ ┌─────────┐
                │ Driver A│ │ Driver B│ │ Driver C│
                └─────────┘ └─────────┘ └─────────┘
                      │          │          │
                      ▼          ▼          ▼
                ┌───────────────────────────────────┐
                │        Design Under Test          │
                └───────────────────────────────────┘
```

338. **Given a simple single port RAM as shown below that can either do a read or a write to an address, write a sequence and driver following UVM to test for read and write. Assume that read_enable=1 means read and write_enable=1 means write. Only one of read or write can happen in a single cycle.**

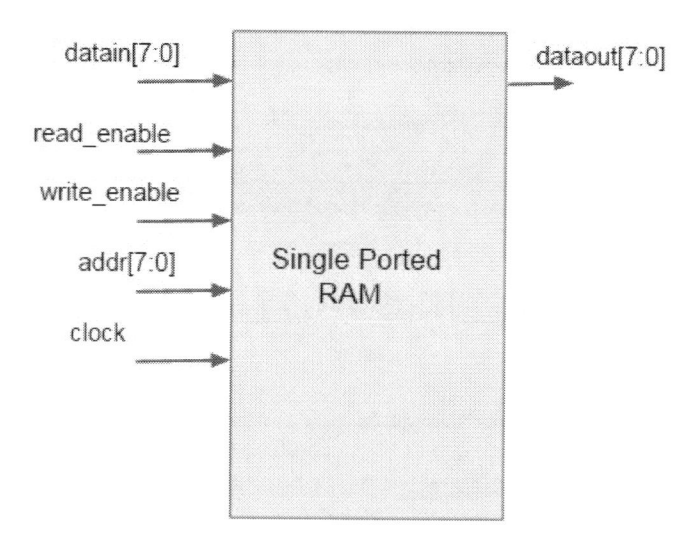

To implement a UVM driver and sequence, we need to first define the sequence item class and then the sequence and driver class can use this as the transaction to communicate. Following is an example code for the sequence item, the sequence and the driver. Use similar guidelines for approaching any programming code using UVM.

1) Sequence item is the transaction used for communication between a sequence and driver. This should abstract all information needed to be finally driven as signals on DUT.

```
class  rw_txn  extends uvm_sequence_item;
  rand bit[7:0] addr;   //address of transaction
  typedef enum {READ, WRITE} kind_e;   //read or write type
  rand kind_e  sram_cmd;
  rand bit[7:0] datain;  //data

  //Register with factory for dynamic creation
  `uvm_object_utils(rw_txn)

  //constructor
  function new (string name = "rw_txn");
    super.new(name);
  endfunction

  //Print utility
  function string convert2string();
    return $psprintf("sram_cmd=%s addr=%0h datain=%0h",sram_cmd.name(),addr,datain);
  endfunction
endclass
```

2) A sequence that generates 10 transactions of above type and send to driver:

```
class sram_sequence extends uvm_sequence#(rw_txn) ;
```

```systemverilog
  //Register with factory
  `uvm_object_utils(sram_sequence)

  function new(string name ="sram_sequence");
    super.new(name);
  endfunction

  //Main Body method that gets executed once sequence is started
  task body();
    rw_txn rw_trans;
    //Create 10 random SRAM read/write transaction and send to driver
      repeat(10) begin
      rw_trans =
rw_txn::type_id::create(.name("rw_trans"),.contxt(get_full_name()));
      start_item(rw_trans); //start arbitration to sequence
      assert (rw_trans.randomize());  //randomize item
      finish_item(rw_trans);   //send to driver
    end
  endtask
endclass
```

3) A driver code that receives above transaction from sequence and drives it on SRAM protocol.

```systemverilog
class sram_driver extends uvm_driver#(rw_txn);
  `uvm_component_utils(sram_driver)

  virtual sram_if vif;  //Interface that groups dut signals

  function new(string name,uvm_component parent = null);
    super.new(name,parent);
  endfunction

  //Build Phase
  //Get the virtual interface handle from config_db

  function void build_phase(uvm_phase phase);
    super.build_phase(phase);
    if (!uvm_config_db#(virtual sram_if)::get(this, "", "sram_if", vif)) begin
      `uvm_fatal("SRAM/DRV", "No virtual interface specified")
    end
  endfunction

  //Run Phase
  //Implement the Driver-Sequencer API to get an item
  //Based on if it is Read/Write - drive on SRAM interface the corresponding pins
  virtual task run_phase(uvm_phase phase);
    super.run_phase(phase);
    this.vif.read_enable  <= '0;
    this.vif.write_enable <= '0;
```

```
    forever begin
      rw_txn tr;
      @ (this.vif.master_cb);
      //First get an item from sequencer
      seq_item_port.get_next_item(tr);
      @ (this.vif.master_cb);  //wait for a clock edge
      uvm_report_info("SRAM_DRIVER ", $psprintf("Got Transaction
%s",tr.convert2string()));
      //Decode the SRAM Command and call either the read/write function
      case (tr.sram_cmd)
        rw_txn::READ:  drive_read(tr.addr, tr.dataout);
        rw_txn::WRITE: drive_write(tr.addr, tr.datain);
      endcase
      //Handshake DONE back to sequencer
      seq_item_port.item_done();
    end
  endtask: run_phase

  //Drive the SRAM signals needed for a Read
  virtual protected task drive_read(input  bit   [31:0] addr,
                                    output logic [31:0] data);
    this.vif.master_cb.addr         <= addr;
    this.vif.master_cb.write_enable <= '0;
    this.vif.master_cb.read_enable  <= '1;
    @ (this.vif.master_cb);
    this.vif.master_cb.read_enable <= '0;
    data = this.vif.master_cb.dataout;
  endtask: drive_read

  //Drive the SRAM signals needed for a Write
  virtual protected task drive_write(input bit [31:0] addr,
                                     input bit [31:0] data);
    this.vif.master_cb.addr         <= addr;
    this.vif.master_cb.write_enable <= '1;
    this.vif.master_cb.read_enable  <= '0;
    @ (this.vif.master_cb);
    this.vif.master_cb.write_enable <= '0;
  endtask: drive_write
endclass
```

339. What is a factory?

A "factory" in UVM methodology is a special look up table in which all of the UVM components and transactions are registered. The recommended way to create objects of components and transactions in UVM is by using the factory method called `create()`. Creating objects using factory also helps in substituting an object of one type with an object of a derived type without having to change the structure of the testbench or editing the testbench code.

340. What is the difference between creating an object using new() and create() methods?

The recommended method in UVM for creating components or transaction objects is to use the built-in method *::type_id::create()* instead of calling the constructor *new()* directly. The create method internally makes a call to the factory to look up the requested type and then calls the constructor new() to actually create an object. This allows type overriding easily as in the test, you can specify the type of class (base or one or derived) and all the other testbench components will be able to create object of that class type without any code change.

A `new()` constructor will only create an object of a given type and therefore using a new() will not allow run time changing of class types. Hence, using a new() means the testbench code will need to change based on the different types to be used.

341. How do we register an uvm_component class and uvm_sequence class with factory?

The uvm_sequence class is registered with the factory using `uvm_object_utils()` macro and passing the class name as argument. An Example below:
```
class  test_seq_c extends uvm_sequence;
    `uvm_object_utils(test_seq_c)
```
The uvm_component class is registered with the factory using `uvm_component_utils()` macro and passing the class name as argument. An Example below:
```
class  test_driver_c extends uvm_component;
    `uvm_component_utils(test_driver_c)
```

342. Why should we register a class with factory?

A factory is a special look up table used in UVM for creating objects of component or transaction types. The benefit of object creation using factory is that a testbench build process can decide at run-time which type of object has to be created. Based on this, a class type could be substituted with another derived class type without any real code change. To ensure this feature and capability, all classes are recommended to be registered with factory. If you do not register a class with factory, then you will not be able to use the factory method `::type_id::create()` to construct an object.

343. What is meant by factory override?

The UVM factory allows a class to be substituted with another derived class at the point of construction. This can be useful for changing the behaviour of a testbench by substituting one class for another without having the need to edit or re-compile the testbench code.

344. What is the difference between instance override and type override?

A **type override** means that every time a component class type is created in a testbench hierarchy, a substitute type is created in its place. This applies to all instances of that component type. On the other hand, an **instance override** means, overriding only a specific instance of a component class. A specific instance of a component is referenced by the position of that component in the UVM component hierarchy. Since only UVM component classes can have a hierarchy in UVM testbenches, only component classes can be overridden on an instance granularity while sequences (or UVM objects) can be only type overridden.

345. Can instance override and type override be used for both UVM_component class and transaction types?

No, only UVM_component classes are part of UVM testbench hierarchy and can be overridden on an instance granularity. The sequence items or sequences are not a part of UVM testbench hierarchy and hence can only be overridden using type override which will override all objects of that type.

346. What is the concept of objections and where are they useful?

The `uvm_objection` class provides a means for sharing a counter between multiple components and sequences. Each component/sequence may "raise" and "drop" objections asynchronously, which increases or decreases the counter value. When the counter reaches zero (from a non-zero value), an "all dropped" condition is said to occur.

The objection mechanism is most commonly used in the UVM phasing mechanism to coordinate the end of each run-time phase. User-processes started in a phase raises an objection first and drops the objection once the process completes. When all processes in a phase drops the objects, the phase's objection count goes to zero. This "all dropped" condition indicates to the phasing mechanism that every participant agrees the phase should be ended.

Following is an example of how a sequence (`my_test_sequence`) is started on a sequencer (`my_sequencer`) and the objection is dropped after sequence completes execution

```
task main_phase( uvm_phase phase);
  phase.raise_objection( this );
  my_test_sequence.start(my_sequencer);
  phase.drop_objection( this );
endtask
```

347. How can we implement a simulation timeout mechanism in UVM methodology?

A simulation time out mechanism helps to stop the simulation if the test doesn't progress because of some bug beyond a maximum time.
In UVM, **set_global_timeout(timeout)** - is a convenience function that sets **uvm_top.phase_timeout** variable with the timeout value. If the run() phase doesn't end before this timeout value, then the simulation is stopped and an error is reported.

This function is called in the top level module which also starts the test as follows

```
module test;
  initial begin
    set_global_timeout(1000ns);
  end

  initial begin
    run_test();
  end
endmodule
```

348. What is the concept of phasing in UVM methodology?

Unlike a module based testbench (in which all modules exist statically in a hierarchy), class based testbench need to manage creation of different objects and execution of various tasks and functions in those objects. Phasing is an important concept in class based testbenches to have a consistent testbench execution flow. A test execution can be conceptually divided into following tasks - configuration, creation of testbench components, runtime stimulus, and end of test checks. UVM defines standard phases for each of these as part of the methodology.

349. What are the different phases of a UVM component? What are the sub-phases for the UVM run_phase()?

UVM uses standard phases to order the major steps that take place during simulation. There are three groups of phases, which are executed in the following order.
1. Build phases - In the build phases; the testbench is configured and constructed. It has following sub-phases which are all implemented as virtual methods in uvm_component base class.
   1) build_phase()
   2) connect_phase()
   3) end_of_elaboration()

2. Run time phases - These phases can consume time and this is where most of the test execution happens.
   1) start_of_simulation()
   2) run_phase()

The run_phase() is further divided into 12 sub-phases as below:
   1) pre_reset

2) reset
3) post_reset
4) pre_configure
5) configure
6) post_configure
7) pre_main
8) main
9) post_main
10) pre_shutdown
11) shutdown
12) post_shutdown

3. Clean up phase - This phase execute after the test ends and is used to collect, and report results and statistics from the test. This consists of following sub phases:
1) extract()
2) check()
3) report()
4) final()

350. Why is build_phase() executed top down in uvm_component hierarchy?

In UVM, all the testbench components like test, Env, Agent, Driver, Sequencer are based of uvm_component class and there is always a hierarchy for the testbench components. The build_phase() method is part of uvm_component class and is used to construct all the child components from the parent component. So, to build the testbench hierarchy you always need to have a parent object first, which can then construct its children, and that can further construct its children using build_phase. Hence, build_phase() is always executed top down. For Example: The top level uvm_test class calls build_phase which should construct all the uvm_env components part of this test, and the build_phase() of each uvm_env class should construct all the uvm_agent components that are part of that uvm_env, and this goes on. For all other phases it really doesn't matter in which order it is called. The run_phase() for all components runs in parallel.

351. What is the use of phase_ready_to_end() method in a uvm_component class?

`phase_ready_to_end(uvm_phase phase)` is a callback method available in a component class which gets called when all objections are dropped for that corresponding phase and the phase is going to end.

A component class can use this callback method to define any functionality that it needs to perform when the phase is about to end.

One example is if a component want to delay ending of phase until some condition even after all objections are dropped, it can be done using this callback method.

Another example is if an irritator or reactive sequence is running until a main sequence is complete, phase_ready_to_end() callback method in main_phase() can be used to stop those irritator sequences.

352. What is uvm_config_db and what is it used for?

The UVM configuration mechanism supports sharing of configurations and parameters across different testbench components. This is enabled using a configuration database called **uvm_config_db**. Any testbench component can populate the configuration database with variables, parameters, object handles etc. Other testbench components can get access to these variables, parameters, object handles from the configuration database without really knowing where it exists in the hierarchy.

For Example, a top level testbench module can store a virtual interface pointer to the uvm_config_db. Then any uvm_driver or a uvm_monitor components can query the uvm_config_db to get handle to this virtual interface and use it for actually accessing the signals through the interface.

353. How do we use the get() and set() methods of uvm_config_db?

The get() and set() are the primary methods used to populate or retrieve information from the uvm_config_db. Any verification component can use the **set()** method to populate the config_db with some configuration information and can also control which other components will have visibility to same information. It could be set to have global visibility or visible only to one or more specific testbench components. The get() function checks for a shared configuration from the database matching the parameters.

The syntax for the get() and set() methods are as follows:

```
uvm_config_db#(<type>)::set(uvm_component context, string inst_name,
string field_name,<type> value)
```

```
uvm_config_db#(<type>)::get(uvm_component context, string inst_name,
string field_name, ref value)
```

The *context* specifies the current class or component from which get/set is called. The *inst_name* is the name of the instance of component from which get/set is called. The *field_name* is the name of the object or parameter or variable which is set/get in config_db. The <type> identifies the type of the configuration information set/get in config_db. For object handles, this will have the class name while for other variables; it will be the type of that variable.

354. Is it possible for a component lower in testbench hierarchy to pass a handle to a component in higher level of hierarchy using get/set config methods?

This is not a recommended way of passing configuration objects in UVM. Normally the higher level component sets up configuration data base with handles and the lower level components do get them using get/set methods.

355. What is the recommended way of assigning virtual interfaces to different components in a UVM verification methodology?

The top level testbench **module** which instantiates the DUT and interfaces will set the virtual interface in the **uvm_config_db**. A test class or any other component in the UVM component hierarchy can then query the uvm_config_db using the get() method to get handles to this virtual interface and use them for accessing signals. Following shows an example of how this is done. The module test actually instantiates a DUT and physical interface for an APB bus master. It then sets the virtual interface handle to the uvm_config_db.

```
module test;
  logic pclk;
  logic [31:0] paddr;
  //Instantiate an APB bus master DUT
  apb_master apb_master(.pclk(pclk),*);
  //Instantiate a physical interface for APB interface
  apb_if  apb_if(.pclk(pclk), *);
  initial begin
    //Pass this physical interface to test class top
    //which will further pass it down to env->agent->drv/sqr/mon
    uvm_config_db#(virtual apb_if)::set(null, "uvm_test_top", "vif", apb_if);
  end
endmodule
```

Following shows a APB Env class that uses the get() method in uvm_config_db to retrieve the virtual interface that was set in the top level test module.

```
class apb_env  extends uvm_env;
  `uvm_component_utils(apb_env);
  //ENV class will have agent as its sub component
  apb_agent  agt;
```

```
        //virtual interface for APB interface
        virtual apb_if  vif;
        //Build phase - Construct agent and get virtual interface handle
from test and pass it down to agent
        function void build_phase(uvm_phase phase);
          agt = apb_agent::type_id::create("agt", this);
          if (!uvm_config_db#(virtual apb_if)::get(this, "", "vif", vif))
          begin
            `uvm_fatal("config_db_err", "No virtual interface specified
for this env instance")
          end
          uvm_config_db#(virtual apb_if)::set( this, "agt", "vif", vif);
        endfunction: build_phase
      endclass : apb_env
```

356. Explain how simulation ends in UVM methodology?

UVM has a phased execution which consists of a set of build phases, run phases and check phases. The `run()` phase is where the actual test simulation happens and during this phase every component can raise an objection in beginning and hold it until it is done with its activity. Once all components drops the objection, the run() phase completes and then check() phase of all components execute and then the test ends.

This is how a normal simulation ends, but there are also controls on simulation timeouts to terminate the `run()` phase if some component hangs due to a bug in design or testbench. When the `run()` phase starts, a parallel timeout timer is also started. If the timeout timer reaches one of the specified timeout limits before the `run()` phase completes, the `run()` phase will timeout, an error message will be issued and then all phases post `run()` will get executed and test ends after that.

357. What is UVM RAL (UVM Register Abstraction Layer)?

UVM RAL (Register Abstraction Layer) is a feature supported in UVM that helps in verifying the registers in a design as well as in configuration of DUT using an abstract register model. The UVM register model provides a way of tracking the register content of a DUT and a convenience layer for accessing register and memory locations within the DUT. The register model abstraction reflects the structure of the design specification for registers which is a common reference for hardware and software engineers working on the design.
Some other features of RAL include support for both front door and back door initialization of registers and built in functional coverage support.

358. What is UVM Call back?

The `uvm_callback` class is a base class for implementing callbacks, which are typically used to modify or augment component behavior without changing the component class. Typically, the component developer defines an application-specific callback class that extends from this class and defines one or more virtual methods, called as *callback interface*. The methods are used to implement overriding of the component class behavior. One common usage can be to inject an error into a generated packet before the driver sends it to DUT. Following pseudo code shows how this can be implemented.
1) Define the packet class with an error bit
2) Define the driver class that receives this packet from a sequence and sends it to DUT
3) Define a driver callback class derived from the base `uvm_callback` class and add a virtual method which can be used to inject an error or flip a bit in the packet.
4) Register the callback class using `` `uvm_register_cb() `` macro
5) In the run() method of driver that receives and send the packet to the DUT, based on a probability knob, execute the callback to cause a packet corruption

```
class Packet_c;
  byte[4]  src_addr, dst_addr;
  byte[]   data;
  byte[4]  crc;
endclass


//User defined callback class extended from base class
class PktDriver_Cb extends uvm_callback;
  function new (string name = "PktDriver_Cb");
    super.new(name);
  endfunction

  virtual task corrupt_packet (Packet_c pkt);
    //Implement how to corrupt packet
    //example - flip one bit of byte 0 in CRC
    pkt.crc[0][0] = ~pkt.crc[0][0]
  endtask
endclass : PktDriver_Cb

//Main Driver Class
class PktDriver extends uvm_component;
 `uvm_component_utils(PktDriver)
  //Register callback class with driver
 `uvm_register_cb(PktDriver,PktDriver_Cb)

  function new (string name, uvm_component parent=null);
    super.new(name,parent);
```

```
  endfunction

  virtual task run();
    forever begin
      seq_item_port.get_next_item(pkt);
      `uvm_do_callbacks(PktDriver,PktDriver_Cb, corrupt_packet(pkt))
      //other code to derive to DUT etc
    end
  endtask
endclass
```

359. What is uvm_root class?

The **uvm_root** class serves as the implicit top-level and phase controller for all UVM components. Users do not directly instantiate **uvm_root**. The UVM automatically creates a single instance of uvm_root that users can access via the global (uvm_pkg-scope) variable, **uvm_top**.

360. What is the parent class for uvm_test?

uvm_test class is the top level class that a user can implement and there is no explicit parent that is mentioned. However, UVM has a special component called uvm_top and it is assigned as the parent of test class.

6.2 Formal Verification

While simulation is still the major component of any hardware verification project lifecycle, importance of Formal Verification is growing, and overtime it has become a significant part of the overall design process. A dynamic simulation can only increase the confidence regarding the design correctness and it can never be 100% complete. However, formal verification has the ability to comprehensively prove correctness of a design against a specification. The complexity though is in the size of the design and hence appropriate areas/features are normally selected for formal verification.

361. What is Formal Verification?

Formal Verification is a process where we use mathematical modelling to verify a Design implementation meets a specification. It uses mathematical reasoning and algorithms to prove that a design meets a specification. In formal verification, all cases (inputs and state) are covered implicitly by the tool without the need for developing any stimulus generators or expected outputs. A formal description of the specification in terms of properties or higher level model is required by the tool for exhaustively covering all input combinations to prove or disprove functional correctness. SystemVerilog properties can be used to formally describe a design specification.

362. Is Formal Verification a Static Verification process or a Dynamic Verification process?

Formal Verification is a Static Verification process as there are no dynamic simulation cycles that are run.

363. What are the different methods for performing Formal Verification?

There are two commonly used formal verification methods:
1) Model Checking
2) Formal Equivalence

364. Describe Model Checking.

In Model Checking method, a model to be verified is described as set of properties that are derived from the design specification. Here, state space of the design is searched exhaustively to see if all the properties hold under all the states. It throws an error if a property is violated for any state. The diagram below describes this:

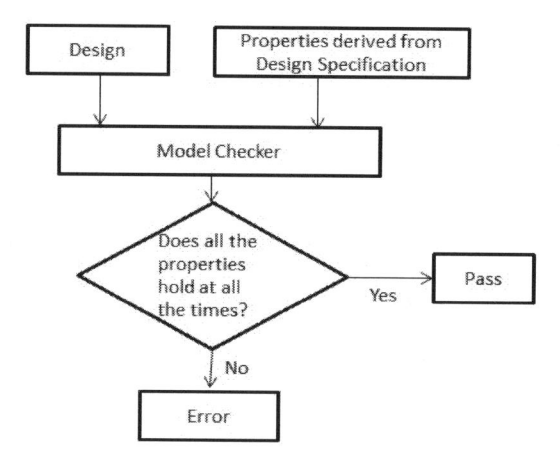

365. Describe Formal Equivalence.

Formal Equivalence is used to verify if two models at same or different abstraction are functionally the same or not. This method doesn't tell if a model is functionally correct, but it tells if two models are functionally the same or not (functionally equivalent). This is most commonly used in comparing functionality of the RTL design and the synthesized netlist. It can also be used to check against two RTL models or two Gate level models. The diagram below represents this:

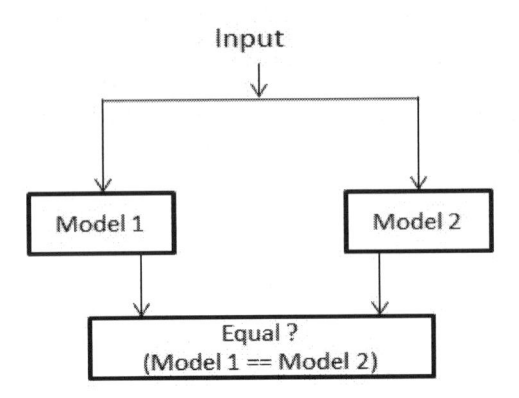

```
                    Input
                      │
           ┌──────────┴──────────┐
           ▼                     ▼
      ┌─────────┐           ┌─────────┐
      │ Model 1 │           │ Model 2 │
      └────┬────┘           └────┬────┘
           │                     │
           ▼                     ▼
        ┌────────────────────────────┐
        │         Equal ?            │
        │   (Model 1 == Model 2)     │
        └────────────────────────────┘
```

366. List down few verification conditions under which you can use Formal Equivalence.

Formal Equivalence can be used to verify if following models are equivalent:
1) RTL Design & Synthesized Netlist (Gate level model)
2) RTL Design & Reference Model
3) Two RTL designs
4) Two Gate level models
5) Two reference models

367. What are the advantages of Formal Verification over Dynamic Verification?

Following are some of the advantages of Formal Verification over Dynamic simulations:

1) Exhaustive verification is not possible with dynamic simulations as the input stimulus is implemented using a generator or tests. However, Formal verification covers exhaustive state space as the stimulus is generated automatically by the tool to try and prove/disprove all specifications.
2) There is no need to generate input stimulus, since exhaustive stimulus is generated automatically by tool. The effort from the user will be to implement a formal specification using properties.
3) There is no need to generate expected output sequences and the correctness of design is guaranteed mathematically.

368. What are limitations of Formal Verification?

Following are some of the limitations of Formal Verification:

1) Scalability is one of the biggest limitations of Formal Verification. Formal Verification is limited to smaller designs because even addition of one flip-flop increases design state space by a factor of 2 (which means input scenarios are doubled for every flip-flop).
2) It ensures correctness of design with respect to design specification. It doesn't guarantee if the design works properly (say if Specification itself is buggy).
3) For model checking, design specification needs to be translated and coded in terms of properties.

369. If a module in a design is formally verified to be working properly, do we need coverage data for that module?

No, we don't need coverage data for a module which is formally verified to be working properly. This is because formal verification mathematically guarantees that the design intent would be verified under all possible input conditions.

6.3 Power and Clocking

Present day low-power ASIC/SoC designs support Power Gating, Dynamic Voltage Frequency Scaling, Multi-Voltage Domains, and Multi-Power domains. Hence, Power Aware Verification and clock domain crossing verification forms an important part of design verification flow. This section introduces basic concepts related to Power and Clocking that you should be aware of.

370. What are the main components of the power consumption in a CMOS circuit?

There are two main components that constitute the power consumed by a CMOS circuit:
1) Dynamic Power - This is caused by transitions because of capacitive charging
2) Static Power - This is caused by leakage currents in the absence of any switching. i.e in idle state.

371. What is Dynamic Power and what all parameters does Dynamic Power consumption depend upon?

Dynamic Power consumption (P_D) is the sum of capacitive load power (P_{cap}) and transient power consumption ($P_{Transient}$). Dynamic power consumption (P_D) is directly proportional to CV^2f.

$$P_D = P_{cap} + P_{Transient}$$
$$P_D = A(C_L+C)V_{dd}^2 f$$

Where, A = Activity Factor (Fraction of the circuit that is switching)

C = Internal Capacitance
C_L = Load Capacitance
V_dd = Supply Voltage
f = Operating Frequency

372. What is Static Power and what all parameters does Static Power consumption depend upon?

Static Power is the power which is consumed when there is no switching activity in the circuit. It is the leakage power. For Example: when you turn on the circuit, battery starts draining because of current flow.

$$P_{static} = I_{static} * V_{dd}$$

Where, V_{dd} is the supply voltage and I_{static} is the total current flowing through the device.

373. Explain the concept of Multi-Voltage domains and why is it needed?

Multi-Voltage domain is a Low-Power Design technique where we have multiple voltage domains in a design. Here, the aim is to optimize power for a required performance. Higher the voltage, faster can be the circuit (higher performance), but higher will be the power consumption as well (since dynamic power is proportional to CV^2f). In some designs, only few portions of the design may need to operate at higher frequency and other portions may run at a lower frequency also. In such cases, less voltage is supplied to those portions of the design that can operate at lower frequency as well. This results in saving power consumption.

374. What is "Dynamic Voltage Frequency Scaling" (DVFS) and when is it used?

Dynamic Voltage Frequency Scaling is a Low-Power Design Technique to dynamically adjust power for a required frequency. In this technique, operating frequency and/or voltage are modified in a way that minimum frequency and/or voltage are needed for proper operation of a design. It is called "Dynamic" as this process happens when the design is operational. Traditional power saving was run at operating frequency and then power off when idle for a periodic scheduling. For the same workload if we are trading performance by running at lower frequency we save much more power. The frequency we chose based on the performance needs to complete a task. This saving also due to the voltage is quadratic. This is done during the run time based on the performance needs for the task optimising the idle to run longer but at lower voltage. This helps in saving power.

375. What is UPF?

UPF stands for Unified Power Format and it is an IEEE 1801 Standard. It is used for specifying power intent of the design. For instance, it describes: which power rails should be routed, when different blocks of design are expected to be powered up/powered down, how voltage levels change as signals moves from one power domain to another, isolation cells and corresponding clamp values, memory retention, etc.

376. What is Power Aware Simulation and why is it important?

Power Aware Simulation means modelling Power Down and Power Up behavior at the RTL and/or GLS level. Power Aware Simulation is important because:
1) It is necessary to find power related RTL/Spec bugs early in the design cycle. Critical power related bugs can lead to chip being non-functional.
2) Power Management is critical and modern day ASIC/SoC designs have significant logic implemented for Power Management.

377. What is meant by a Power Domain?

A Power Domain is a collection of design elements that share a primary supply set. These are the portions of the design that are grouped based on common power strategy like operating voltage, supply nets, power up/down conditions, etc.

378. What is the need of having level shifters in a design?

Level shifters are required when multiple voltage domains are present in a design. When a logic signal moves from one voltage domain to another voltage domain, level shifter is used to convert that logic signal from one voltage level to another voltage level so that signal logic value is interpreted properly in different voltage domains.

379. Define the concept of Isolation with respect to Power Domains.

Isolation is a technique for controlling behavior of a signal which is driven into or out of a power domain that is powered down. It comprises of clamping the signal to a known value i.e. 1, 0, or latching it to a previous value when the power domain is powered down.

380. What is meant by metastability? When does it happen and what are its consequences?

Metastability (Metastable state) is a state where a circuit is not able to settle at a stable "0" or "1" logic level within the time required for proper operation of the circuit. This usually happens when there are setup and hold time violations. Following are its consequences:
1) It can lead to unpredictable system behavior.
2) Different fan-out cones can read different values of the signal and can cause the design to enter an unknown state.
3) If unstable data ("0" or "1") propagates to different portions in the design, it can lead to high current and can eventually result in chip burn-out.

381. How can metastability be avoided?

Metastability can be avoided by using synchronizers in the design. Synchronizers ensure sufficient time for the unstable oscillations ("0" and "1") to settle down such that a stable output is obtained.

382. How does a basic synchronizer circuit look like? Draw a sample Synchronizer circuit.

Below is an example of a basic synchronizer circuit. It is a two flip-flop synchronizer. The first flip-flop waits for one clock cycle to allow for any metastability at the input to settle down/fade away and then second flip-flop provides a stable signal at the output.

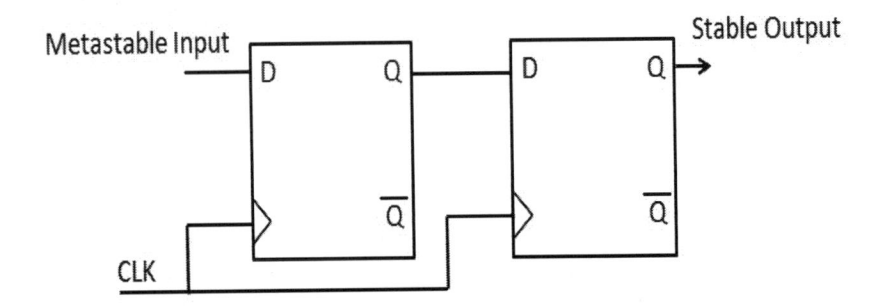

However, it is still possible for the output of first flip-flop to be unstable at the time signal is clocked into second stage (and cause stage-2 output signal to go metastable). In such scenarios, we can use three flip-flop synchronizer circuit. Having said this, usually two flip-flop synchronizer circuits are sufficient to remove metastability.

383. What is Clock Gating?

Clock Gating is a power saving technique where supply of clock to inactivate portions of the design is shut-off. It is a technique that is used to control power dissipated by the clock network. This helps in reducing dynamic power consumed by the design by avoiding unnecessary switching activity.

384. What is Power Gating and why is it used?

Power Gating is a power saving technique in which portions of the design that are not in use are shut-down. Power gating shuts off static leakage when a logic is not in use, thereby reducing power consumption. Power Gating is a low-power design technique. Clock gating helps in reducing dynamic power while power gating helps in reducing static power.

385. What kind of issues can be encountered in designs (having multiple Clock Domains) at Clock Domain Crossings (CDC)?

Following are the most common type of issues that can be encountered at a Clock Domain Crossing (CDC). A Verification Engineer should be familiar with these scenarios so as to better verify a design with multiple clock domains:

1) **Metastability leading to synchronization failure in the design**: Clocks are running at different frequencies in different clock domains and when a signal generated in one clock domain is sampled very close to the active edge of clock in second clock domain, output may go to metastable state leading to synchronization failure in the design.

2) **Data incoherency**: Destination clock domain may receive incoherent data if proper design techniques are not used. For Example: If multiple signals are being transferred from one clock domain to another clock domain such that all these signals are changing simultaneously, and the source and destination active clock edges arrive close to each other, some of these signals may get captured in one clock cycle and some in another clock cycle in destination clock domain, thereby leading to data incoherency. **Note**: This is just one example of data-incoherency. There can be many more scenarios where data-incoherency may happen.

3) **Data Loss**: Data may be lost at the CDC boundary if proper design techniques are not used. For Example: If a signal is moving from a faster clock domain to a slower clock domain and the width of that signal is only equal to one clock period (of faster running clock), there can be a case where this information is missed if the signal arrives between the active clock edges of clock in slower clock domain. **Note**: This is just one example of data-loss. There can be many more scenarios where data-loss may happen.

386. How can you synchronize signal(s) between two clock domains? Mention few design techniques that can be used.

Design techniques to synchronize signal(s) between two clock domains would differ depending upon whether we need to pass 1-bit or multiple-bits between different clock domains. As explained in one previous answer, case of multiple CDC bits should be handled carefully. Assume a case where multiple signals (multiple CDC bits) are being transferred from one clock domain to another clock domain such that all these signals are changing simultaneously, and the source and destination active clock edges arrive close to each other. In such a case, some of these signals may get captured in one clock cycle and some in another clock cycle in destination clock domain leading to data incoherency.
Following design techniques can be used to synchronize signal(s) between two clock domains.

For 1-bit crossing clock domain:
1) Use 2 Flip-flop or 3 Flip-flop synchronizer in the receiver(destination) clock domain. Concept of synchronizer is discussed in one of the answers above.
2) Use closed Loop synchronizer. In this technique, feedback signal is sent as an acknowledgement signal from destination clock domain to source clock domain.

For multiple-bits crossing clock domain:

1) Formation of Multi-Cycle Paths (MCP) for multiple CDC bits. In this method, unsynchronized signal from source clock domain is sent to destination clock domain along-with a synchronized control signal.
2) Gray Code encoding for multiple CDC bits. As gray code allows only 1-bit change, it can tackle metastability issues.
3) Using Asynchronous FIFOs (data-in from source clock domain and data-out from destination clock domain) for multiple CDC bits. This is **one of the safest design** techniques as it provides full synchronization independent of clock frequency.
4) Consolidating multiple CDC bits into 1-bit CDC signals and using 2 Flip-flop synchronizer or closed loop synchronizer techniques with 1-bit CDC signal.

387. Give an example of an issue that can occur while transferring data from a faster clock domain to a slower clock domain.

If a signal is moving from a faster clock domain to a slower clock domain and the width of that signal is only equal to one clock period (of faster running clock), there can be a case where this information is missed if the signal arrives between the active clock edges of clock in slower clock domain.

388. What are the advantages and disadvantages of using an Asynchronous Reset?

Advantages:
1) Asynchronous resets get highest priority.
2) Data path is guaranteed to be clean.
3) They can occur with or without the presence of a clock signal.

Disadvantages:
1) If Asynchronous reset is de-asserted at (or near) an active edge of the clock, output of a flip-flop may go into a metastable state.
2) It is sensitive to glitches, and this may lead to spurious resets.

389. What are the advantages and disadvantages of using a Synchronous Reset?

Advantages:
1) Synchronous resets occur at an active clock edge and hence they ensure that a circuit is 100% synchronous.
2) They are easier to work with while using cycle based simulators.
3) They usually synthesize into smaller flip-flops thereby helping save die-area.

Disadvantages:
1) Synchronous resets may need a pulse-stretcher so that it's wide enough to be seen at active clock edge.
2) They lead to addition of extra combinational logic.

3) Synchronous resets need presence of clock to cause the resets. If the circuit has an internal tristate bus, separate asynchronous reset would be required to prevent bus conflict on internal tristate bus.

390. What is a Reset Recovery Time? Why is it relevant?

Reset Recovery Time is the time between reset de-assertion and the clock signal going high. If the reset de-assertion happens and within a very small time window, if the clock signal goes high, it can lead to metastability conditions. This is because all signals that become active after reset de-assertion will not meet timing conditions at the next flop input.

391. What is a frequency synthesizer? Give an example of frequency synthesizer?

Frequency synthesizer is a circuit that can generate a new frequency from a single stable reference frequency. For Example: For generating a 200MHz clock signal from a reference 100 MHz clock signal, **PLL** is commonly used as a frequency synthesizer.

392. What is a PLL?

PLL stands for "Phase Locked Loop". In simple terms, it is a feedback electronic circuit (a control system to be precise) that is used to generate an output signal whose phase is related to the phase of an input signal. It is used for performing phase/frequency modulation and demodulation, and can also be used as frequency synthesizer. PLL consists of three functional blocks:
1) Phase Detector
2) A Loop Filter
3) Voltage Controlled Oscillator

393. Draw a block diagram depicting the use of PLL as a frequency synthesizer?

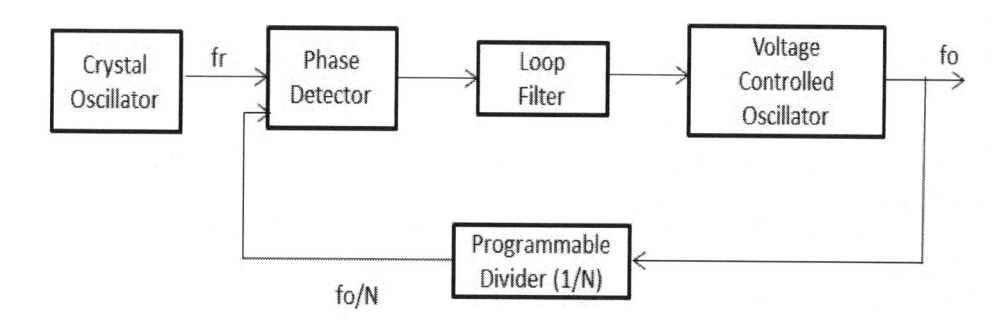

Here fr is the reference frequency and fo is the output frequency such that `fr = fo/N`, which implies that `fo = N*fr`

6.4 Coverage

Coverage is defined as the percentage of verification objectives that have been met. It is used as a metric for evaluating the progress of a verification project. Coverage metric forms an important part of measuring progress in constrained random testbenches and also provides good feedback to the quality and effectiveness of constrained random testbenches. Broadly there are two types of coverage metrics - Code Coverage and Functional Coverage. While code coverage is generated automatically by simulators, Functional coverage is user defined and normally implemented using constructs supported by SystemVerilog language. This section has questions related to the coverage concepts as well as the SystemVerilog language constructs used for implementing functional coverage model.

394. What is the difference between code coverage and functional coverage?

There are two types of coverage metrics commonly used in Functional Verification to measure the completeness and efficiency of verification process.

1) **Code Coverage:** Code coverage is a metric used to measure the degree to which the design code (HDL model) is tested by a given test suite. Code coverage is automatically extracted by the simulator when enabled.

2) **Functional Coverage:** Functional coverage is a user-defined metric that measures how much of the design specification, as enumerated by features in the test plan, has been exercised. It can be used to measure whether interesting scenarios, corner cases, specification invariants, or other applicable design conditions — captured as features of the test plan — have been observed, validated, and tested. It is user-defined and not automatically inferred. It is also not dependent on the design code as it is implemented based on design specification.

395. What are the different types of code coverage?

Code coverage is a metric that measures how well the HDL code has been exercised by the test suite. Based on the different program constructs, code coverage are of following types:
1) **Statement/Line coverage**: This measures how many statements (lines) are covered during simulation of tests. This is generally considered important and is targeted to be 100% covered for verification closure. In the following example code, you can see there are 4 lines or statements which will be measure in statement/line coverage.
```
always @ (posedge clk) begin
  if( A > B)  begin    //Line 1
     Result = A - B;   //Line 2
```

```
    end else begin       //Line 3
      Result = A + B;    //Line 4
    end
  end
```

2) **Block coverage**: A group of statements between a begin-end or if-else or case statement or while loop or for loop is called a block. Block coverage measures whether these types of block codes are covered during simulation. Block coverage looks similar to statement coverage with the difference being that block coverage looks for coverage on a group of statements. In the same example code as shown below you can see there are three blocks of code (Enclosed in 3 begin ... end)

```
always @ (posedge clk) begin   //always block
  if( A > B)   begin     // if block
    Result = A - B;
  end else begin         // else block
    Result = A + B;
  end
end
```

3) **Branch/Decision coverage**: Branch/Decision coverage evaluates conditions like if-else, case statements and the ternary operator (?:) statements in the HDL code and measures if both true and false cases are covered. In the same example above there is a single branch (if A >B) and the true and false conditions will be measured in this type of coverage.

4) **Conditional Coverage and Expression coverage**: Conditional coverage looks at all Boolean expressions in the HDL and counts the number of times the expression was true or false. Expression coverage looks at the right-hand side of an assignment, evaluates all the possible cases as a truth table and measures how well those cases are covered. Following is an expression of 3 boolean variables that can cause the `Result` variable to be true of false

$$Result = (A \&\& B) \;||\; (C)$$

You can create a truth table as follows for all possible cases of A, B and C that can cause result to be true or false. The expression coverage gives a measure of if all the rows of this truth table are covered.

A	B	C	Result
0	0	0	0
0	0	0	0
0	0	1	1
0	1	0	0
0	1	1	1
1	0	0	0
1	0	1	1
1	1	0	1
1	1	1	1

5) **Toggle coverage:** Toggle coverage measures how well the signals and ports in the design are toggled during the simulation run. It will also help in identifying any unused signals that does not change value.

6) **FSM coverage:** FSM coverage measures whether all of the states and all possible transitions or arcs in a given state machine are covered during a simulation.

396. During a project if we observe high functional coverage (close to 100%) and low code coverage (say < 60%) what can be inferred?

Remember that code coverage is automatically extracted by simulator based on a test suite while functional coverage is a user defined metric. A low code coverage number shows that not all portions of the design code are tested well. A high functional coverage number shows that all the functionalities as captured by the user from the test plan are tested well. If coverage metric shows low code coverage and a high functional coverage then one or more of following possibilities could be the cause:
1) There could be lot of design code which are not used for the functionality implemented as per design specification. (also known and dead code)
2) There is some error in the user defined functional coverage metrics. Either the test plan is not capturing all the design features/scenarios/corner cases or implementation of functional coverage monitors for same is missing. The design code not covered in code coverage could be mapping to this functionality.
3) There could be potential bugs in implementing functional coverage monitors causing them to be showing as falsely covered. Hence it is important in the verification project to have proper reviews of the user defined functional coverage metrics and its implementation.

397. During a project if we observe high code coverage (close to 100%) and low functional coverage (say < 60%) what can be inferred?

If coverage metric shows high code coverage and a low functional coverage then one or more of following possibilities could be the cause:
1) Not all functionality is implemented in the design as per the specification. Hence the code for same is missing while functional coverage metrics exists with no test
2) There could be potential bugs in the functional coverage monitor implementation causing them to be not covered even though tests might be simulated and exercising the design code.
3) There could be a possibility that tests and stimulus exists for covering all features but those might be failing because of some bugs and hence being excluded from the measurement for functional coverage.

398. What are the two different types of constructs used for functional coverage implementation in SystemVerilog?

SystemVerilog language supports two types of implementation - one using covergroups and the other one using cover properties.

covergroups: A covergroup construct is used to measure the number of times a specified value or a set of values happens for a given signal or an expression during simulation. A covergroup can also be useful to measure simultaneous occurrence of different events or values through cross coverage. Similar to a class, once defined, a covergroup instance can be created using the new() operator. A covergroup can be defined in a package, module, program, interface, checker, or a class.

cover-properties: A cover property construct can be used to measure occurrences of a sequence of events across time. This uses the same temporal syntax used for creating properties and sequences used for writing assertions.

399. Can covergroups be defined and used inside classes?

Yes, covergroups can be defined inside classes and it is a very useful way to measure coverage on class properties. This is useful to implement functional coverage based on testbench constructs like transactions, sequences, checkers, monitors, etc.

400. What are coverpoints and bins?

A coverage point (**coverpoint**) is a construct used to specify an integral expression that needs to be covered. A **covergroup** can have multiple coverage points to cover different expressions or variables. Each coverage point also includes a set of bins which are the different sampled values of that expression or variable. The bins can be explicitly defined by the user or automatically created by language.

In the following example there are two variables **a** and **b** and the covergroup has two coverpoints that look for values of **a** and **b** covered.

The coverpoint cp_a is user defined and the bins values_a looks for specific values of **a** that are covered.

The coverpoint cp_b is automatic and the bins are automatically derived which will be all the possible values of b.

```
  bit [2:0] a;
  bit [3:0] b;
  covergroup cg @(posedge clk);
    cp_a : coverpoint a {
      bins values_a = { [0,1,3,5,7] };
    }
    cp_b : coverpoint b;
  endgroup
```

401. How many bins are created in following examples for coverpoint cp_a ?
```
        bit[3:0] var_a;
        covergroup  test_cg @(posedge clk);
          cp_a : coverpoint var_a {
```

```
            bins low_bins[] = {[0:3]};
            bins med_bins   = {[4:12]};
       }
    endgroup
```

Four bins are created for `low_bins[ ]` where each of the bin look for the specific value from 0 to 3 for coverage hit separately. One bin is created for `med_bins` which will look for any value between 4 and 12 for it to be covered.
So, total 5 bins are created for the coverpoint `cp_a`.

402. What is the difference between ignore bins and illegal bins?

ignore_bins are used to specify a set of values or transitions associated with a coverage point that can be explicitly excluded from coverage. For example, following will ignore all sampled values of 7 and 8 for the variable a.

```
coverpoint a {
   ignore_bins ignore_vals = {7,8};
}
```

illegal_bins are used to specify a set of values or transitions associated with a coverage point that can be marked as illegal. For example, following will mark all sampled values of 1, 2, 3 as illegal.

```
covergroup cg3;
   coverpoint b {
      illegal_bins bad_vals = {1,2,3};
   }
endgroup
```

If an illegal value or transition occurs, a runtime error is issued. Illegal bins take precedence over any other bins, that is: they will result in a run-time error even if they are also included in another bin.

403. How can we write a coverpoint to look for transition coverage on an expression?

Transition coverage is specified as "value1 => value2" where value1 and value2 are the sampled values of the expression on two successive sample points. For example, below coverpoint looks for transition of variable v_a for values of 4, 5 and 6 in three successive positive edges of clk.

```
covergroup cg @(posedge clk);
   coverpoint v_a {
      bins sa = (4 => 5 => 6);
   }
endgroup
```

404. What are the transitions covered by following coverpoint?
```
coverpoint my_variable {
  bins trans_bin[] = ( a,b,c => x, y);
}
```

This will expand to cover for all of the transitions as below:
a=>x, a=>y, b=>x, b=>y, c=>x, c=>y

405. What does following bin try to cover?
```
covergroup test_cg @(posedge clk);
  coverpoint var_a {
    bin  hit_bin = { 3[*4]};
  }
endgroup
```

The **[*N]** is a consecutive repetition operation. Hence, above bin is trying to cover a transition of the signal var_a for 4 consecutive values of 3 across successive sample points (positive edge of clk).

406. What are wildcard bins?

By default, a value or transition bin definition can specify 4-state values. When a bin definition includes an X or Z, it indicates that the bin count should only be incremented when the sampled value has an X or Z in the same bit positions, i.e., the comparison is done using ===. A wildcard bin definition causes all X, Z, or ? to be treated as wildcards for 0 or 1.
For example:
```
coverpoint a[3:0] {
  wildcard bins  bin_12_to_15 = { 4'b11?? };
}
```
In the above bin_12_to_15, lower two bits are don't care and hence if sampled value is any of 1100, 1101, 1110 or 1111, then the bin counts same.

407. What is cross coverage? When is cross coverage useful in Verification?

A coverage group can specify cross coverage between two or more coverage points or variables. Cross coverage is specified using the `cross` construct. Cross coverage of a set of **N** coverage points is defined as the coverage of all combinations of all bins associated with the **N** coverage points which is same as the Cartesian product of the **N** sets of coverage point bins.
```
bit [31:0] a_var;
bit [3:0] b_var;
covergroup cov3 @(posedge clk);
  cp_a:  coverpoint a_var {
    bins yy[] = { [0:9] };
```

```
    }
    cp_b:    coverpoint b_var;
    cc_a_b : cross cp_b, cp_a;
endgroup
```

In the above example, we define a cross coverage between coverpoints cp_a and cp_b. The cp_a will have 10 bins that look for values from 0 to 9 while cp_b will have 16 bins as b_var is a 4 bit variable. The cross coverage will have 16*10 = 160 bins.
A cross coverage can also be defined between a coverpoint and a variable in which case an implicit coverpoint will be defined for that variable.
Cross coverage is allowed only between coverage points defined within the same coverage group.
Cross coverage is useful in verification to make sure that multiple events or sample values of expressions are happening simultaneously. This is very useful because a lot of time it is important to test the design for different features or scenarios or events happening together and cross coverage helps to make sure all those combinations are verified.

408.   How many bins are created by following cross coverage?
```
bit[1:0] cmd;
bit[3:0] sub_cmd;
covergroup abc_cg @(posedge clk);
  a_cp: coverpoint cmd;
  cmd_x_sub: cross cmd, sub_cmd;
endgroup
```

The a_cp will generate 4 automatic bins as cmd is a 2 bit signals. Crossing a coverpoint with a variable will cause SystemVerilog to create an implicit coverpoint for the variable. Hence, an implicit coverpoint for sub_cmd will be created which will have 16 automatic bins. Therefore, the cross coverage will generate 4*16=**64 bins**

409.   What can be wrong with following coverage code?
```
int   var_a;
covergroup  test_cg @(posedge clk);
  cp_a: coverpoint var_a {
    bins low = {0,1};
    bins other[] = default;
  }
endgroup
```

This covergroup has a coverpoint that tries to cover the sampled values of an integer which can be 2^{32} values. The first bin looks for two values 0 and 1, while the second bin that uses default creates a separate bin for all other values (2^{32}-2). This can cause a simulator to crash as the number of bins is huge. The usage of default should be avoided as much or used with care.

410. What are the different ways in which a covergroup can be sampled?

A covergroup can be sampled in two different ways:
1) **By specifying a clocking event with the covergroup definition:** If a clocking event is specified then all the coverage points defined in the covergroup are sampled when the event is triggered. For example in the code below, the clocking event is defined as posedge of the clock (clk). So the covergroup is sampled on every positive edge of clock and the coverpoints are evaluated and counted.
```
covergroup   transaction_cg  @(posedge clk)
   coverpoint  req_type;
endgroup
```

2) **By explicitly calling the predefined sample() method of covergroup:** Sometimes you would not want the covergroup to be sampled on every clock edge or any general event that happens frequently. This is because the expression or variable that you are sampling may not be changing very frequently. In this case, the predefined method `sample()` can be called explicitly when any of the signals or expressions in the covergroup changes. This is a useful way when covergroups are defined inside classes. For example in the following reference code, the covergroup (pkt_cg) is defined inside a class and instantiated inside the constructor. In the test module, the covergroup sample() method is called each time a new packet is created to measure the coverage.
```
class  packet;
  byte[4] dest_addr;
  byte[4] src_addr;

  covergroup pkt_cg;
    coverpoint dest_addr;
  endgroup

  function new();
    pkt_cg =new();
  endfunction;
endpacket

module test;
  initial begin
    packet pkt =new();
    pkt.pkt_cg.sample();
  end
endmodule
```

411. How can we pass arguments to a covergroup and when will that be useful?

Covergroups can include arguments using syntax similar to tasks and functions. Signals can be passed to the covergroup using the "ref" keyword as shown in the example below. Usage of arguments in covergroup is very useful when we want to define a covergroup and reuse the same for multiple instances but with different signals/variables being passed.

For example in the below module test, there is a xy_cg that takes two reference signals and creates a generic covergroup. The example further shows creating two instances of this covergroup by passing signals from two different modules.

```
module test;
  covergroup xy_cg ( ref int x , ref int y, input string name);
    cp_x: coverpoint x;
    cp_y: coverpoint y;
    cc_x_y: cross cp_x, cp_y;
  endgroup

  initial begin
    xy_cg xy_cg_mod1 = new( top.mod1.x, top.mod1.y, "mod1_cvg");
    xy_cg xy_cg_mod2 = new( top.mod2.x, top.mod2.y, "mod2_cvg");
  end
endmodule
```

412. Can coverpoints inside a covergroup reference hierarchical signals in a design?

Yes, coverpoints inside a covergroup can reference signals in design using the hierarchy.

413. Can we have cross coverage between coverpoints in different covergroups?

No, cross coverage is only possible for coverpoints in same covergroup.

414. What is the difference between coverage per instance and per type? How do we control the same using coverage options?

A covergroup can be defined and instantiated multiple times. If there are multiple instances of a covergroup, then by default SystemVerilog reports the coverage for that group as cumulative coverage across all instances. This default behavior is coverage per covergroup type. However, there is a per_instance option that can be set inside a covergroup and then SystemVerilog will report coverage separately for each instance of the covergroup.

```
covergroup test_cg @(posedge clk)
  option.per_instance =1;
  coverpoint var_a;
  //and other coverpoints
endgroup
```

6.5 Assertions

An assertion specifies a behavior of the system. Assertions are primarily used to validate the behavior of a design. In addition, assertions can also be used to provide functional coverage, and to flag that input stimulus, which is used for validation. Assertions can be checked dynamically by simulation, or statically by property checking or formal verification tools.

SystemVerilog supports rich constructs to implement assertions in terms of sequences and property specifications. This section will cover commonly asked questions related to SystemVerilog Assertions and methodology and help you understand these better.

415. What is an assertion and what are the benefits of using assertions in Verification?

An assertion is a description of a property of the design as per specification and is used to validate the behavior of the design. If the property that is being checked for in a simulation does not behave as per specification, then the assertion fails. Similarly if a property or rule is forbidden from happening in the design and occurs during simulation, then also the assertion fails.

Following are some of the benefits of using Assertions in Verification:
1) Assertions improve error detection in terms of catching simulation errors as soon a design specification is violated
2) Assertions provide better observability into design and hence help in easier debug of test failures.
3) Assertions can be used for both dynamic simulations as well as in formal verification of design
4) Assertions can also be used to provide functional coverage on input stimulus and to validate that a design property is infact simulated.

416. What are different types of assertions?

There are two types of assertions defined by SystemVerilog language - immediate assertions and concurrent assertions.

417. What are the differences between Immediate and Concurrent assertions?

Immediate assertions use expressions and are executed like a statement in a procedural block. They are not temporal in nature and are evaluated immediately when executed. Immediate assertions are used only in dynamic simulations. Following is an example of a simple immediate assertion that checks "if a and b are always equal":

```
always_comb   begin
  a_eq_b: assert (a==b) else $error ("A not equal b");
end
```

Concurrent assertions are temporal in nature and the test expression is evaluated at clock edges based on the sampled values of the variables involved. They are executed concurrently with other design blocks. They can be placed inside a module or an interface. Concurrent assertions can be used with both dynamic simulations as well static (formal) verification. Following is a simple example of a concurrent assertion that checks "if c is high on a clock cycle, then on next cycle, value of a and b is equal":

```
ap_a_eq_b : assert property((@posedge clk) c |=> (a == b));
```

418. What is the difference between simple immediate assertion and deferred immediate assertions?

Deferred assertions are a special type of immediate assertions. Simple immediate assertions are evaluated immediately without waiting for variables in its combinatorial expression to settle down. Hence, simple immediate assertions are very prone to glitches as the combinatorial expression settles down. This can cause the assertions to fire multiple times and some of them could be false.

To avoid this, deferred assertions are defined which gets evaluated only at the end of time stamp when the variables in the combinatorial expression settles down. This means they are evaluated in the reactive region on the timestamp.

419. What are the advantages of writing a checker using SVA (SystemVerilog Assertions) as compared to writing it using a procedural SystemVerilog code?

Certain types of checkers are better written using SVA rather than procedural code. The languages supports rich constructs to implement sequence and property specifications and this becomes easier than using procedural code or writing class based checkers. The other added benefit is that the same assertions can also be used in static checking tools like a Formal Verification tool as well as in providing functional coverage.

Some examples where SVA can be used better are following:
1) Checking of internal design structures like FIFO's overflowing or underflowing.
2) Checking of internal signals and interfaces between modules can be easier done with embedded assertions in design
3) Checkers for standard interface protocols like PCIE, AMBA, Ethernet, etc. can also be easily developed using the temporal expressions.
4) Checks for arbitration, resource starvation, protocol deadlocks, etc. are normally candidates for Formal Verification in any design and hence writing assertions for these will help them to be used in both static and dynamic simulations.

420. What are the different ways to write assertions for a design unit?

1) Assertions can be written directly inside a design module. This is mostly followed if the assertions are written by design engineers for some of the internal signals or interfaces in the design.
2) Assertions can also be written in a separate interface or module or program and then that can be bound to a specific module or instance from which signals are referenced in assertion. This is done using the bind construct in SystemVerilog. This method is generally followed if the assertions are written by the Verification engineers.

421. What is a sequence as used in writing SystemVerilog Assertions?

A sequence is a basic building block for writing properties or assertions. A sequence can be thought of a simple boolean expression that gets evaluated on a single clock edge or it can be a sequence of events that gets evaluated across multiple cycles. A property may involve checking of one or more sequential behaviors beginning at various times. A property can hence be constructed using multiple sequences combined logically or sequentially. The basic syntax of a sequence is as follows:

```
sequence name_of_sequence;
  <boolean expression >
endsequence
```

For Example: The following sequence samples values of a and b on every positive edge of clk and evaluates to true if both a and b are equal.

```
sequence  s_a_eq_b;
  @posedge(clk)  (a ==b);
endsequence
```

422. Is there a difference between $rose(tst_signal) and @posedge(tst_signal)?

Yes, there is a difference. @posedge(*tst_signal*) waits until a rising edge event is seen on the *tst_signal*. However, $rose() is a system function that checks if the sampled value of the signal changed to 1 between previous sample and the current sample (Previous sample could be a 0/x/z). Accordingly, $rose(tst_signal) only returns true if there are at least two sampled values.

For example: In the following sequence, only if the signal "a" changes from a value of 0/x/z to 1 between two positive edge of clock, then $rose(a) will evaluate true

```
sequence S1;
  @(posedge clk) $rose(a);
endsequence
```

423. When does following sequence evaluate to true?

```
sequence S1;
  @(posedge clk)  $rose(a);
endsequence
```

1) When the signal "a" changes from 0 to 1.
2) When the signal "a" had a value of "0" at one posedge of clk which changes to "1" at the next posedge of clk.
3) When the signal "a" had a value of "1" at one posedge of clk and "0" at the next posedge of clock.

2). The $rose() system function evaluates to true if the value changes from 0 to 1 when sampled on two consecutive clock cycles as explained in previous question.

424. Can a sequence be declared in?
1) Module
2) Interface
3) Program
4) Clocking Block
5) Package

Yes, a sequence can be declared in any of the above.

425. Is it possible to have concurrent assertions implemented inside a class?

No, concurrent assertions cannot be implemented inside a class

426. Explain when the following sequence matches?
```
req ##2 gnt ##1 !req
```

When *gnt* signal goes high two cycles after *req* signal is high, and one cycle after that *req* signal is deasserted to zero, this sequence will evaluate to true.

427. What is a sequence repetition operator? What are the three different type of repetition operators used in sequences?

If a sequential expression needs to be evaluated for more than one iteration, then instead of writing a long sequence, repetition operator can be used to construct a longer sequence. SVA supports three types of repetition operators:

1) **Consecutive Repetition** ([*const_or_range_expression]): If a sequence repeats for a finite number of iterations with a delay of one clock tick from end of one iteration, then a consecutive repetition operator can be used. Following is an example of how to use consecutive repetition operator.
```
a ##1 b [*5]
```

In above example, if "a" goes high and then if "b" remains high for 5 consecutive cycles, we can use the repetition operator [*] to specify number of iterations.

2) **Go-to repetition** ([->const_or_range_expression]): Go-to repetition specifies finitely many iterative matches of the operand Boolean expression, with a delay of one or more clock ticks from one match of the operand to the next successive match and no match of the operand strictly in between. The overall repetition sequence matches at the last iterative match of the operand.

   ```
   a ##1 b [->2:10] ##1 c
   ```

   In the above example, the sequence matches over an interval of consecutive clock ticks provided **a** is true on the first clock tick, **c** is true on the last clock tick, **b** is true on the penultimate clock tick, and, including the penultimate, there are at least 2 and at most 10 not necessarily consecutive clock ticks strictly in between the first and last on which **b** is true.

3) **Non-consecutive repetition** [=const_or_range_expression]): The Non-consecutive repetition is like the Go-to repetition except that a match does not have to end at the last iterative match of the operand Boolean expression

   ```
   a ##1 b [=2:10] ##1 c
   ```

   Above sequence shows the same example using non-consecutive repetition. The difference between this repetition and the Go-to repetition is that in this case: after we see a minimum of 2 and maximum of 10 occurrences of non-consecutive **b,** there can be several cycles where **b** is not true and then **c** can be true. Whereas, in a sequence that uses Go-to repetition, after the maximum number of **b** occurrences are seen, next cycle needs to have **c** as true.

428. Find any issue (not syntax errors) with following assertion?

```
module test (input clk, input a, input b);
   assert_1: assert ( a && b);
endmodule;
```

Immediate assertions can be started only inside procedural blocks

429. Write an assertion check to make sure that a signal is high for a minimum of 2 cycles and a maximum of 6 cycles.

Following property uses a sequence such that if a signal "a" rises, then from same cycle, we check it remains high for a minimum of 2 and maximum of 6 cycles and in the next cycle "a" goes low.

```
property a_min_2_max_6: @(posedge clk)
   $rose(a) |-> a[*2:6] ##1 (a==0)
endproperty
assert property (a_min_2_max_6);
```

430. What is an implication operator?

An implication operator specifies that the checking of a property is performed conditionally on the match of a sequential antecedent. This construct is used to precondition monitoring of a property expression and is allowed only at the property level. Following is the syntax of two types of implication operators supported in property expressions:

1) **Overlapped Implication Operator** (|->)
   ```
   assert property  prop_name  ( sequence_expr |-> property_expr )
   ```

2) **Non-Overlapped Implication Operator** (|=>)
   ```
   assert property  prop_name  ( sequence_expr |=> property_expr )
   ```

In above examples, the left hand side of the implication operator is called antecedent and the right hand side of the operator is called consequent. The antecedent is the precondition that needs to happen before evaluating the consequent.

431. What is the difference between an overlapping and nonoverlapping implication operator?

Overlapped Implication Operator (|->): For overlapped implication, if there is a match for the antecedent sequence_expr, then the endpoint of the match is the start point of the evaluation of the consequent property expression. For Example: In following example, as soon as a match happens on the sequence (a==1), in the same cycle if "b" is true and the following cycle "c" is true then this property passes.
```
assert property  abc_overlap (@posedge clk  (a==1)  |-> b ##1 c )
```

Non-Overlapped Implication Operator (|=>): For non overlapped implication, the start point of the evaluation of the consequent property_expr is the clock tick after the end point of the match of antecedent. For Example: In following example, when (a==1) matches on any clock cycle, then in next cycle if "b" is true and a cycle later if "c" is true, then following property will pass.
```
assert property  abc_overlap  (@posedge clk  (a==1)  |=> b ##1 c  )
```

432. Can implication operator be used in sequences?

No, it can be used only in properties. It is a precondition match to evaluate property expressions.

433. Are following assertions equivalent?
```
1) @(posedge clk) req |=> ##2 $rose(ack);
2) @(posedge clk) req |-> ##3 $rose(ack);
```

Yes: |-> is an overlapping operator that starts evaluating the consequent in same cycle when antecedent is true while |=> is non overlapping operator that starts consequent evaluation a cycle after antecedent is true. So, adding an explicit cycle delay after overlapping operator will make it equivalent to non-overlapping operator.

434. Is nested implication allowed in SVA?

Yes. These are useful when we have multiple gating conditions leading to a single final consequence. For Example: a |=> b |=> c
Here, when "a" is true, then next cycle "b" is evaluated and then if found true, next cycle "c" is evaluated, and if found true, the property passes.

435. What does the system task $past() do?

$past is a system task that is capable of getting values of signals from previous clock cycles.

436. Write an assertion checker to make sure that an output signal never goes X?

The system function $isunknown(signal) returns a value of 1 if the signal has an unknown value (x). Hence this can be used to write an assertion as below.
```
assert property (@(posedge clk) ! $isunknown(mysignal));
```

437. Write an assertion to make sure that the state variable in a state machine is always one hot value.

The $isonehot() system function returns if a bit vector is one hot. Hence, this can be used to write an assertion as follows:
```
assert property (@(posedge clk)  $isonehot(state));
```

438. Write an assertion to make sure that a 5-bit grant signal only has one bit set at any time? (only one req granted at a time)

The system function $countones() will return the number of ones present in a signal. Hence, this can be used to write an assertion to check for number of bits set in any signal as follows:
```
assert property (@(posedge clk) $countones(grant[5:0])==1);
```

439. Write an assertion which checks that once a valid request is asserted by the master, the arbiter provides a grant within 2 to 5 clock cycles

```
property  p_req_grant;
  @(posedge clk)  $rose (req) |->  ##[2:5] $rose (gnt);
endproperty
```

440. How can you disable an assertion during active reset time?

A property can use a "**disable iff**" construct to explicitly disable an assertion. Following is an example that disables an assertion check when reset is active high.
```
assert property (@(posedge clk) disable iff (reset) a |=> b);
```

441. What's the difference between assert and assume directives in SystemVerilog?

An **assert** directive is used to specify the property as an obligation for the design that is to be checked to verify that the property holds.
An **assume** directive is same as assert in simulation. It is used to specify the property as an assumption for the environment. Simulators check that the property holds, while formal tools use the assume directive as a constraint information to generate input stimulus.

442. What is bind construct used in SystemVerilog for?

The bind construct in SystemVerilog is used to externally instantiate (or bind) a module or interface or checker to a target module or an instance of module. This is useful for any instrumentation code or assertions that are encapsulated in a module, interface, program, or checker to be instantiated in a target module or a module instance without modifying the target module code.
The syntax is:
```
bind  <target module/instance>  <module/interface to be instantiated> <instance name with port map>
```

For example, following code shows an interface named "range" which has an assertion implemented as shown below

```
interface range (input clk, enable, int minval, expr);
  property crange_en;
    @(posedge clk) enable |-> (minval <= expr);
  endproperty
  range_chk: assert property (crange_en);
endinterface
```

1) This could be instantiated (bind) inside a module - say called as **cr_unit** as shown below. Effectively every instance of module **cr_unit** will also have an instance of this interface (r1)
```
bind cr_unit range r1(c_clk,c_en,v_low,(in1&&in2));
```

2) If we want to instantiate (bind) the interface only with a very specific instance of the module cr_unit (let's say cr_unit_1), then we can use following as example:
```
bind cr_unit:cr_unit_1 range r1(c_clk,c_en,v_low,(in1&&in2));
```

443. How can all assertions be turned off during simulation?

Assertions can be turned off during a simulation using the $assertoff() system task. If no arguments are specified, all the assertions are disabled.
If this system task is called in the middle of simulation, then any active assertions at that given point of time are allowed to complete before disabling.
For selectively disabling assertions, the task supports two arguments as follows:
```
$assertoff[(levels[, list])]
```
The first argument specifies how many levels of hierarchy this applies and the second argument is a list of properties that need to be turned off in these levels of hierarchy.

444. What are the different ways in which a clock can be specified to a property used for assertion?

There are different ways in which a clock can be specified to a property as explained below:
1) A sequence instance that is used in property has an explicit clock specified. In this case property uses that clock.
   ```
   sequence seq1;
     @(posedge clk) a ##1 b;
   endsequence
   property prop1;
     not seq1;
   endproperty
   assert property (prop1);
   ```

2) Specify the clock explicitly in the property.
   ```
   property prop1;
     @(posedge clk) not (a ##1 b);
   endproperty
   assert property (prop1);
   ```

3) Infer the clock from the procedural block in which the property is used as shown below.
   ```
   always @(posedge clk) assert property (not (a ##1 b));
   ```

4) If the property is defined in a clocking block, the clock of the clocking block can be inferred in the property. The property can be used to assert outside by hierarchical reference as shown below:
   ```
   clocking master_clk @(posedge clk);
     property prop1;
       not (a ##1 b);
     endproperty
   endclocking
   assert property (master_clk.prop1);
   ```

5) If none of the above is used, then the clock will be resolved to the default clocking event. For Example: if a clocking block (shown above) has defined a default clocking event (as shown below) then the property infers the same clock.
```
default clocking master_clk ; // master clock as defined above
property p4;
   not (a ##1 b);
endproperty
assert property (p4);
```

445. **For a synchronous FIFO of depth=32, write an assertion for following scenarios. Assume a clock signal (clk), write and read enable signals, full flag and a word counter signal.**
  1) **If the word count is >31, FIFO full flag is set.**
  2) **If the word count is 31 and a new write operation happens without a simultaneous read, then the FIFO full flag gets set.**

```
assert property (@(posedge clk) disable iff (!rst_n) (wordcnt>31 |-> fifo_full));
assert property (@(posedge clk) disable iff (!rst_n) (wordcnt==31 && write_en && !read_en |=> fifo_full));
```

Note that for the second case, a non-overlapping implication operator is used as the full flag will go high only in next cycle after write_enable is seen.

Chapter 7: Version Control Systems

Version Control Systems have been an integral part of Software Engineering domain for a long-long time. But now they are slowly gaining popularity in Hardware Engineering domain as well. With Hardware Designs becoming more and more complex, various new design features getting integrated every quarter, multiple folks working on the same database across different sites, version control systems have become indispensable. Hence, this section touches upon basics of various Version Control Systems.

After a few general questions, we look into basic commands for three most popular Version Control Systems: CVS, GIT and SVN. This section aims at making our readers familiar with different version control systems and hence, we are providing basic commands which can help the readers refresh their knowledge (applicable to readers who have worked on these systems before), and get started with these systems (applicable to readers who are new to Version Control Systems).

Note: There are lot of similarities in various commands used in different version control systems.

7.1 General

446. What is a Version Control System?

A Version Control System is a database that stores all the change records of your work.

447. What is the need of a Version Control System?

When multiple members of a team work together on a shared project, it is important to keep incremental changes of all individual team members in sync in a common database. A Version Control System helps in achieving this by updating author's incremental changes to a common database with author's name, changes made, and their comments. This information can be accessed by other people whenever required.

448. What are some examples of Version Control Systems?

CVS (Concurrent Version Systems), GIT, SVN (Subversion), Perforce, etc.

449. What is a repository?

A repository is a central place where all the data/code is stored and maintained. It's a central storage where all the files and directories (which are part of a project) are stored.

7.2 CVS

450. What is CVS?

CVS stands for "Concurrent Versions System" and it is a commonly used version control system which is available for free.

451. How to add a new file or directory in the CVS database?

cvs add <filename>
This command adds the specified <filename> in the CVS database. Once this filename is added, you need to "checkin" the file so that the actual file contents are updated in the database.

452. How to check-in a file to a CVS database?

After adding a filename, you can put a file into the database using following commands.
cvs checkin -m "message_here" <filename>, OR
cvs ci -m "message_here" <filename>, OR
cvs commit -m "message_here" <filename>

Where, –m "message_here": is a message option to specify any short information about the change. You can check-in as many as versions of a file and each check-in gives you a new version number.

453. How to check out a file from CVS Database?

cvs checkout <filename>, OR
cvs co <filename>

Here, the <filename> can be a file or a directory (if you want to check-out all files in a directory). By default above command will always get you the latest version of the file. If you want to get a specific version of a file, you need to use the version number as following:
cvs co -r <version_number> <filename>

454. How to update files in your working environment to the latest in CVS database?

Once you have checked out a database or a set of files/directories, there may be other users who could modify the same or different files in the central database, and you would want these changes to reflect in your workspace at regular intervals.

For this, you need to use the "cvs update" command which works in a way similar to checkout command. Following command updates all the file/files which were updated in the CVS but were not synced with your working environment.
cvs update < filename>, OR
cvs up <filename>

455. What is tagging and how to tag a file?

Tagging is a very useful feature provided by version control systems. While working on a shared project database with a group of people, you might want to add checkpoints to the database at regular intervals, (For Example: when a project reaches intermediate milestones). This is possible by associating a tag with all the files (in a database) that might have different versions.
You can provide a same tag to all the relevant files from different versions and you can retrieve them anytime by supplying the tag name. To tag a file, following command is used:
cvs tag <filename>

456. How to check-out a set of files (in a module) from a given tag?

If you want to get all the files from the repository with a given tag, you can use following command:
cvs co -r <tag> <module_name>

457. How can you delete a tag?

Following command is used to delete tag information from a file:
cvs rtag -d <tagname> <filename>

458. How to find difference between files from two different versions?

If you want to know the difference between the files from two different CVS versions, you can use following command:
cvs diff -r <version1> -r <version2> <filename>

459. How can you see the check-in log messages for a file?

Log messages, which were given at the time of check-in with -m option, can be seen by using:
cvs log <filename>

460. How can you check the status of a file?

To know the status of a file i.e. to check if a file is in sync with the file in the central database or if it has been modified locally, use:
cvs status <filename>

461. How can you view the tag information with the status of a file?

To view the tag information with the status, use following command:
cvs status -v <filename>

7.3 GIT

462. What is GIT?

GIT is one of the most widely used open source version control system both for software development as well as hardware design. It is also available for free and is a distributed revision control system.

463. What's the advantage of using GIT over CVS?

CVS only tracks changes to single files whereas GIT tracks entire source trees with a global view.

464. What would git command "git init" do?

It would Initialize/Create a GIT repository.

465. What git command is used to create a git repo for your personal changes/development?

git clone <repository_to_be_cloned>

466. Which command is used to fetch the latest updates from other repositories?

git pull <repository_from_which_updates_are_to_be_fetched>

467. Which command is used to publish your changes to the group?

git push

468. Where does git keep track of what version you have checked out?

GIT keeps track of what version you have checked out in HEAD. HEAD points to a branch which contains a SHA1 hash. Following git command is used to find out this information:
git head

469. How do you rename a file in GIT?

It is a two-step process:
git mv <filename> <new_filename>
git commit <new_filename>

470. What command is used to view the history of commits to a file or directory?

git log <filename>

471. What command is used to see line-by-line details regarding who changed a file?

git blame <filename>

472. What command is used to show differences between commits or branches?

git diff <commit1> <commit2>

473. What git command is used to undo changes made to a file in your local repository?

git reset <filename>

474. How can you temporarily save changes before pulling/merging or switching branches?

git stash

475. What git command is used to move your changes since your last commit to the staging area?

git add <filename(s)>

476. What git command is used to store the saved changes in the repository and add a message "first commit"?

git commit -m "first commit"

477. How do you revert a commit that has already been pushed and made public?

git revert HEAD~2..HEAD

7.4 SVN

478. What is SVN?

SVN stands for "Subversion" and it is a open source version control system.

479. What is "branch" , "Tag" and "Trunk" in SVN?

Trunk is the main body of development, originating from the start of the project till the end.
Branch is a copy of code derived from a certain point in the trunk that is used for applying major changes to the code while preserving the integrity of the code in the trunk.
Tag is a point in time on the trunk or a branch that you wish to preserve. This is like baselining the code after a major release.

480. What is the difference between Update and Commit?

Update is used to update your local workspace with the changes committed by the team to the repository whereas Commit is the process to push changes from your local area to repository.

481. What is the SVN command to add a file or dir?

svn add <file_or_directory_name>

482. What is the command to create a new directory in SVN?

svn mkdir <new_directory_name>

483. What is the command to view the difference between the local version and repository version of a file?

svn diff <filename>

484. What does the result codes G and R in SVN indicates?

G code: Changes on the repo were automatically merged into the working copy.
R code: This code indicates that item has been replaced in your working copy. This means the file was programmed or scheduled for deletion, and a new file with the same name was scheduled for addition in its place.

485. What is the command to create a new tag from the trunk?

svn copy http://example.svm.com.../repo/trunk http://example.svm.com.../repo/tags/new_tag -m "creating a new tag from trunk"

486. What is the function of Revert in subversion?

Revert function will remove your local changes and reload the latest version from the repository.
Command: svn revert <filename>

Chapter 8: Logical Reasoning/Puzzles

Logical Reasoning Questions and Puzzles form an important part of an interview process. These are aimed towards checking aptitude and logical problem solving skills of an interviewee. For VLSI Verification interviews, these can be broadly classified into three categories:
1) Related to Digital Logic,
2) General Reasoning, and
3) Lateral Thinking.

In most of the cases, interviewers usually don't worry too much about the final answer, but what is really looked at is the approach you take to solve a problem. Hence, it's important to explain your approach and thought process to an interviewer. In this section, we have tried to list down and cover different types of puzzles through limited number of questions that can give you a good background and flair of puzzles asked. To highlight the approach taken to solve these puzzles, we have provided detailed explanations in the answers.

8.1 Related to Digital Logic

For Digital VLSI verification interviews, puzzles relating to Digital Logic are of prime importance as they help recruiters test your digital logic skills as well as your aptitude and your thinking/reasoning ability.

487. Implement following digital gates ONLY using mathematical operations i.e. ONLY using +, -, *
  1) Single Input NOT Gate (input A, output X)
  2) Dual Input AND Gate (inputs A, B, output X)
  3) Dual Input OR Gate (inputs A, B, output X)

1) X = 1 - A (Can be easily driven from truth table of NOT gate)
2) X = A*B   (Simple multiplication)
3) X = A + B - A*B
   From truth table of OR gate, we can say that: X' = (A'.B')

   => X = (A'.B')'
   => X = [(1-A).(1-B)]' (NOT gate: A' = 1-A, as shown in the first part of this answer)
   => X = [(1-A)*(1-B)]' (AND gate A.B = A*B, as shown is the second part of this answer)
   => X = [1 - (1-A)*(1-B)] = A + B - A*B

488. You have 100 coins laying flat on a table. Each coin has two sides: a tail and a head. 10 of them are placed heads up and 90 are placed tails up.

> You can't feel, see or in any other way find out which side is up. Now, split the coins into two piles such that there are equal numbers of heads up in each pile.

Make two piles of the coins. Put 10 coins in one pile and 90 coins in another. Flip all the coins in Pile-1 (the pile with 10 coins). Now, both the piles have equal number of heads up. (Note that there were a total of 10 out of 100 coins placed with heads up and 90 with tails up)

To get more clarity on this answer, assume that Pile-1 you made had only 3 coins with heads up and hence remaining 7 coins with heads up went to Pile-2. If you flip all the coins in Pile-1, Pile-1 will also have 7 coins with heads up.

489. You are given eight identical looking balls. One of them is heavier than the rest of the seven (all the others weigh exactly the same). You are provided with a simple mechanical balance to weigh the ball. What is the minimum number of trials required to figure out the heavier ball?

Two trials will be needed.
Divide the balls in three groups. Group-1 with three balls, Group-2 with three balls and Group-3 with two balls.
In the first trial, put three balls belonging to Group-1 on one side of the balance and three balls belonging to Group-2 on other side of the balance. Now there are two possibilities:
Possibility 1: Both sides are equal. This would mean that heavier ball belongs to Group-3 (group with two remaining balls). If this is the case, in second trial: put one ball each from Group-3 on either side of the balance and you would know the heavier ball.
Possibility 2: Both sides are not equal. From the mechanical balance, you would be able to figure out the Group having heavier ball. Now, in the second trial: randomly pick up two balls from the heavier group and place them on either side of the mechanical balance. Again, if both the sides are equal, you know that 3rd ball in the same group is heavier. Else, you would be able to see heavier ball from the balance.

490. Suppose there are 4 prisoners named W, X, Y, and Z. Prisoner W is standing on one side of a wall, and prisoners X Y and Z are standing on the other side of the wall. Prisoners X, Y, and Z are all standing in a straight line facing right – so X can see prisoner Y and Z, and Y can see prisoner Z. This is what their arrangement looks like:

W || X Y Z

Where, the "||" represents a wall. The wall has no mirrors. So, prisoner W can see the wall and nothing else. There are 2 white hats and 2 black hats and each prisoner has a hat on his head. Each prisoner cannot see the color of his own hat, and cannot remove the hat from his own head. But the prisoners do know that there are 2 white hats and 2 black hats amongst themselves. The prison guard says that if one of the prisoners can correctly guess the color of his hat then the prisoners will be set free and released. The puzzle for you is to figure out which prisoner

would know the color of his own hat?
Note that the prisoners are not allowed to signal to each other, nor speak to each other to give each other hints. But, they can all hear each other if one of them tries to answer the question. Also, you can assume that every prisoner thinks logically and knows that the other prisoners think logically as well.

Prisoner X or Prisoner Y.

As mentioned in the puzzle, Prisoner X can see the hats on the head of Y and Z. If Y and Z have hats of same color on their head (either both white or both black), X would know that in such a situation, he would be having a hat of different color on his head and he would answer the color of the hat on his head correctly.

Now, if X doesn't answer the question for some time, Y would infer that since X is not able to answer the question, Y and Z must be having hats of different color on their heads. Hence, Y would be able to answer the question after looking at color of hat on Z's head. If Z has a white hat on his head, Y would answer black, else white.

491. There are five persons. Out of these five, only one is the truth teller and the remaining four are togglers i.e. that they may tell the truth or may lie on being asked a question. But on being asked again, they will switch i.e. if they told a lie the first time, they will tell the truth on second question and vice versa. You need to ask only two questions to determine who the truth teller is. You can ask both the questions from the same person or ask one question each from two different people. How will you determine who is the truth teller?

Pick any one person and ask first question: "Are you the truth teller?". Now, there can be two responses to this question: "Yes" or "No".

If the response to this question is Yes, then that person can be either a truth teller or a toggler who is lying. In such a scenario, ask the same person second question: "Who is the truth teller?" If the picked person is the truth teller, his response to second question would be "I am the truth teller". Otherwise, if the picked person is a toggler who lied to first question, he would have to tell truth for the second question and point to the truth teller.

Else if the response to first question is No, then that person is a toggler who is telling the truth. Now, since the toggler has told the truth once, he would lie to second question. Ask second question to this person: "Who is not the truth teller?", and you will find out the truth teller.

492. There are 100 prisoners in 100 different prisons. There is a bulb in each prison which is controlled by a switch outside that prison. Initially bulbs in the all the prisons are glowing. In the first iteration, jailor toggles the bulb switches for each and every prison (1, 2, 3, 4, ..., 100). In the

second iteration, jailor toggles the bulb switches only for every 2nd prison (2, 4, 6, 8, ..., 100). In the third iteration, jailor toggles the bulb switches for every 3rd prison (3, 6, 9, 12, ..., 99). Jailor repeats this exercise 100 times where in the 100th iteration he toggles bulb switch for every 100th prison (100). What all prisons will have bulb switched OFF after 100 such iterations.

1, 4, 9, 16, 25, 36, 49, 64, 81, 100 (Squares of a number are odd multiples)
Explanation: Initially all prisons have bulb ON and for the bulb to be OFF, it should go through odd number of iterations. We can either answer this question directly (if we know the fact that ONLY squares of a number are odd multiples), or we can reach to this conclusion by writing down result of 100 iterations for first few prisons/bulbs (say 10). If we right down the results of first 10 prisons:
Prison 1: Would toggle in first iteration only. Hence, this would turn OFF and would never toggle to ON again.
Prison 2: Would get toggled twice (Iteration 1 and Iteration 2). Would restore ON state.
Prison 3: Would get toggled twice (Iteration 1 and Iteration 3). Would restore ON state.
Prison 4: Would get toggled thrice (Iteration 1, 2 and 4). Would hence turn OFF.
Prison 5: Would get toggled twice (Iteration 1 and Iteration 5). Would restore ON state.
Prison 6: Would get toggled four times (Iteration 1, 2, 3 and 6). Would restore ON state.
Prison 7: Would get toggled twice (Iteration 1 and Iteration 7). Would restore ON state.
Prison 8: Would get toggled four times (Iteration 1, 2, 4 and 8). Would restore ON state.
Prison 9: Would get toggled thrice (Iteration 1, 3 and 9). Would hence turn OFF.
Prison 10: Would get toggled four times (Iteration 1, 2, 5 and 10). Would restore ON state.
Hence, we see that out of first ten prisons, only prison 1, 4 and 9 would have bulb switched off after 100 iterations. This gives hint that squares of a number are the ones with off multiples and hence we can extend this logic.

8.2 General Reasoning

493. Two bulbs of 10W and 100W are connected in series with an AC power supply of 100V. Which bulb will glow brighter and why?

10W bulb.

The bulbs are in series so the amount of current flowing through each is same.
P = V^2/R => R = V^2/P
R1 = 100^2/10 = 1000 ohms for 10 Watt bulb
R2 = 100^2/100 = 100 ohms for 100 Watt bulb

Since both the bulbs are in series, applying Voltage division rule:
Voltage across 10 W bulb = 100*1000/(1000+100)=90.91 V

Voltage across 100 W bulb = 100*100/(1000+100)= 9.09 V

Power for 10 W bulb = V^2/R=90.91^2/1000= 8.26 W
Power for 100 W bulb = V^2/R=9.09^2/100= 0.826 W
Hence, 10W bulb glow brighter

494. There are two doors in front of you. One door leads to heaven and the other door leads to hell. There is a guard in-front of each door. One guard always tells the truth and the other guard always lie, but you don't know which one is honest and which one is liar. Given that you can only ask one question from one of them, what would your question be in order to find the way to heaven?

"If I ask the other Guard about which door leads to heaven, what would he tell me?".

The door that the Guard specifies will lead to hell. Other door would lead to heaven.
This is due to the fact that if you end up asking this question from the Guard who always tells the truth, he would point you to the door that leads to hell as he knows that the other guard would lie and would point you to the door that leads to hell. On the other hand, if you end up asking this question from the Guard who always lies, he would lie and point you to door to hell.

495. You have one person working for you for exactly seven days. You have a gold bar to pay him. The gold bar is segmented into seven connected pieces. You must give the person a piece of gold at the end of every day. You can make only two cuts to the gold bar. Where would be these two cuts to allow you to pay the worker 1/7th gold bar each day?

Make first cut so that you have a part that is1/7th of the total size (comprising of 1 segment) and make the second cut so that you have a part with 2/7th of the total size (comprising of 2 segments). Remaining part would be 4/7th of total size (comprising of 4 segments). Now:

Day 1: Give the worker the first part (1/7th of the gold bar i.e. 1 segment).
Day 2: Give the worker the second part (2/7th of the gold bar i.e. 2 segments) and take back the first (1/7th of the gold bar).
Day 3: Give the worker the first part again (so that he now has 3/7th of the gold bar).
Day 4: Give the worker the third part (4/7th of the gold bar i.e. 4 segments) and take back the first and second (3/7th of the gold bar).
Day 5: Give the worker the first part (so that he now has 5/7th of the gold bar).
Day 6: Give the worker the second part and take back the first (so that he now has 6/7th of the gold bar).
Day 7: Give the worker the first part again (so that he now has the entire bar).

496. Four people (say A, B, C and D) need to cross a shaky bridge at night and they have only one torch with them. It's dangerous to cross the bridge without a touch. The bridge can support only two people at a time. All people take different times to cross the bridge (A = 1 min, B = 2 mins, C = 5 mins, and D = 10 mins). What is the shortest time required for all four of them to cross the bridge?

A and B cross the bridge first with the torch (Time: 2 mins, Total Time: 2 mins)

A leaves B on the other side of the bridge and returns with the torch (Time: 1 min, Total Time: 3 mins)

C and D cross the bridge with the torch (Time: 10 mins, Total Time: 13 mins)

B returns with the torch (Time: 2 mins, Total Time: 15 mins)

Finally, A and B cross to the other side of the bridge again (Time: 2 mins, Total Time: 17 mins)

497.   Find out the next term in the series: F13, S15, T17, T19, S21, M23, __ ?

W25
F11: Friday 13th
S13: Sunday 15th
T15: Tuesday 17th
T17: Thursday 19th
S19: Saturday 21th
M21: Monday 23rd
Next date in the series would be 25th and day would be Wednesday. Hence, the answer would be W25.

498.   Two hundred people line up to board a plane with 200 seats. First person (say Jack) in the line gets into the plane and suddenly can't remember his seat number, and hence he randomly picks a seat. After that, each person entering the plane either sits in their assigned seat (if it is available), or if not, chooses an unoccupied seat randomly. When the 200th passenger (say Jill) finally enters the plane, what is the probability that she finds her assigned seat unoccupied?

0.5 (50%)
Let us assume that there are only two seats in the plane. If Jack sits on his seat, Jill would find her seat unoccupied and if Jack sits on Jill's seat, Jill would find her seat occupied. Hence, 50% chances.
Let us now consider all 200 seats. Let's assume Jack sits on the seat belonging to 35th person in the line. Persons 2 to 34 will sit on their own seats, and when person 35 comes in, he can sit either on the seat belonging to Jack or some random seat. If person 35th sits on Jack's seat, Jill will find her seat. Else if the 35th sits on Jill's seat, Jill won't find her seat unoccupied (50% probability). Else if 35th person sits on some other seat at random (which neither belongs to Jack, nor to Jill), decision would be postponed and would depend on the person whose seat 35th person would occupy.

So basically, probability is always 50%: just that decision of Jill finding her seat occupied/unoccupied gets postponed/delayed based upon whose seat Jack sits on and so on.

499. Two old friends, A and B, meet after a long time.
A: Hi B, How are you?
B: I am good. I got married and now I have three kids.
A: Congrats! How old are your kids?
B: Product of their ages is 72 and the sum of their ages is the same as your birth date.
A: But I still don't know.
B: My eldest kid just started taking piano lessons.
A: Oh now I get it.
How old are Bill's kids?

3, 3 and 8
As the product of the ages of the kids is 72, possible values are:
1) 1,2,36
2) 1,3,24
3) 1,4,18
4) 1,6,12
5) 1,8,9
6) 2,2,18
7) 2,3,12
8) 2,4,9
9) 2,6,6
10) 3,3,8
11) 3,4,6

Since sum of the ages is equal to birth date, it has to be between 1 to 31.
Now, as A is not able to find their age from this data, it means that there are two or more sets with same sum. Now sum is same for only two of the above cases: 2,6,6 and 3,3,8. Since, eldest one is taking piano lessons, answer has to be 3,3,8 as 2,6,6 consists of two eldest sons (which is not possible).

500. There is a triangle and on it are 3 ants, one on each corner, and they are free to move along the sides of the triangle. What is probability that the ants will collide?

It is given that ants can move only along the sides of the triangle in any direction. Let's assume that one direction is represented by 1 and another by 0. Since there are 3 sides: eight combinations are possible (2^3). When all ants are going in same direction, they won't collide: that is 111 or 000. Hence, probability of no collision is: 2/8=1/4, and probability of collision is: 6/8=3/4.

8.3 Lateral Thinking

Lateral Thinking puzzles are sort of hybrid between puzzles and stories. In each puzzle, clues/hints to a specific scenario are given. Puzzle solver needs to fill in the details to complete the scenario. These puzzles are usually inexact and may have more than one possible answer. Depending upon the nature of the puzzle (i.e. completeness of the data given in the puzzle), it's usually acceptable for the puzzle solver to ask few questions from the person asking the puzzle in order to arrive at the solution. Answer to such questions (asked by puzzle solver), can be given by person hosting the puzzle in only "yes" or "no". However, since we don't have an option of presenting these puzzles in an interactive manner, we are sharing few puzzles to make sure that the readers are aware of this category of puzzles. The puzzles mentioned below have enough data for our readers to reach a logical conclusion.

This category of puzzles teaches a candidate to check his/her assumptions about any situation. You need to be creative and open-minded to solve these puzzles.

501. **A man lives on the tenth floor of a building. Every day he takes the elevator to go down to the ground floor to go to work. When he returns, he takes the elevator to the sixth floor and walks up the stairs to reach his apartment on the tenth floor. When there are other people in the lift, he goes to tenth floor directly. He hates walking, so why does he walk from sixth floor to tenth floor every day?**

Man is a dwarf and hence can't reach the buttons for 7th, 8th, 9th and 10th floors (as buttons for these floors are away from his reach)

502. **There are six eggs in the basket. Six people each take one of the eggs. How could it be that one egg is left in the basket?**

The last person took the basket with the last egg still inside.

503. **There is a flash of light and a man dies. How?**

Man is struck by lightning.

504. **A hunter aimed his gun carefully and fired. Seconds later, he realized his mistake. And Minutes later, he was dead.**

It was winter timeframe, and the hunter fired the gun near snowy cliff that started an avalanche.

505.   An avid bird watcher sees an unexpected bird. Soon, they were both dead.

Avid bird watcher is sitting in an aeroplane and sees the bird getting stuck into the engine of aeroplane, leading to a plane crash.

506.   How could a baby fall out of a 27 story building on the ground and live?

The baby fell out of a ground floor window even though the building has 27 floors.

Chapter 9: Non Technical and Behavioral Questions

Behavioral interview forms a key part of hiring process. It's very important to do well in this section. Having strong technical skills alone won't suffice as you are usually expected to work as part of a team. Behavioral skills form an important part of work culture of any company and hence this section is usually taken very seriously by recruiters. Performing great in this section with poor technical skills may not fetch candidate a job, but performing bad in this section can definitely cost candidate a job.

Answer to questions in this section would be individual specific. There are no right or wrong answers for these questions. Since answers would be different for different individuals, it's not possible for us to provide answers for these questions. We are providing these questions to help our readers get familiar with type of questions that may be asked in this round of Interview process. However, Best answers to these questions would be the ones that reflect your actual thought process and are truthful. Being open about your aspirations, strengths and weakness always help. **A tip**: Try avoiding tailor made answers and be honest.

507. Tell me something about yourself.

508. What are your strengths and weaknesses?

509. Why do you want to leave your current job (if applicable)?

510. Why do you want to work for this job you are being interviewed for?

511. What are your short term and long term career goals?

512. Why should we hire you?

513. What kind of work would you want to do if you join us?

514. How would your present Manager/Boss describe you?

515. How do you deal with difficult personalities/peers/subordinates/seniors?

516. If you have to pick up one area of your expertise in a non-technical domain, what would it be?

517. What's your favorite programming language? Why?

518. What all technical area(s)/topic(s) are you MOST comfortable with?

519. Give an example of a time you faced a conflict while working in a team. How did you handle it?

520. We all make mistakes we wish we could take back. Tell about a time you wish you had handled a situation differently with a colleague.

521. Tell about a time you needed to get information from someone who wasn't very responsive. What did you do?

522. Tell about a time you failed. How did you deal with this situation?

523. Tell about a time you were under a lot of pressure. How did you get through it?

524. Sometimes it's not possible to complete everything on the to-do list. Tell about a time your responsibilities got overwhelming. What did you do?

525. Give an example of a time when you were able to successfully persuade someone to see things your way at work.

526. Tell about your proudest professional accomplishment.

527. Describe a time when you saw some problem and took the initiative to correct it rather than waiting for someone else to do it.

528. Are you a leader or a follower?

529. If I call your boss right now and ask him about an area you could improve on, what would he say?

530. Can you explain this gap in your employment/educational history? (If there is a gap)

531. Describe your typical day at work.

532. Tell about a time when you disagreed with your boss.

533. Describe your work-style?

534. Tell about a time when you did something wrong. How did you handle it?

535. Have you ever had a difficulty with a supervisor? How did you resolve the conflict?

536. If there were two things you could have changed about your present job, what would those be?

537. What do you like the most about your present job?

538. Describe a situation when your work was criticized. How did you handle it?

539. Do you have any plans for higher/further studies?

540. Where all (What all organizations) have you applied for a job in addition to our organization? Have you been interviewed somewhere else also?

Closing Remarks

So here we are. We thank you for reading this book. We hope that you would have benefitted from reading this book and it would have helped you test, brush-up, and hone fundamental concepts related to Digital VLSI Verification. To-reiterate, a question-bank can never be 100% complete, however big it may be. This book is our sincere effort to cover as many concepts as we could, through limited set of 500+ questions but covering all concepts that will help you answer more related questions.

Feedback is one of the best way to let us know your thoughts on this book. While a positive feedback from you would: give us a sense of satisfaction, help us know what we have done right, encourage us to continue in the same direction, a negative/developmental feedback: would let us know what we could have done better and help us course-correct in future. Please do leave your reviews and ratings that reflect your honest opinion on this book.

If you liked this book, we request you to refer it to your juniors, peers, fellow friends, seniors and spread a word about it, so that others could also benefit from it.

You can stay in touch with us through different mediums as mentioned in the beginning of this book.

Hopefully, you are now ready for your upcoming interview.

All the Best!
Ramdas M and Robin Garg

Made in the USA
Middletown, DE
28 January 2019